Two Manuals Combined into One Book

Marine Scouting and Patrolling

plus

Military Training for Urban Operations

Department of Defense

Pentagon
Publishing

TABLE OF CONTENTS

PART I. SCOUTING

PART I. SCOUTING

CHAPTER 1. FUNDAMENTALS OF SCOUTING

Scouting involves observing terrain and/or the enemy, and accurately reporting those observations. Scouting requires proficiency in the use of weapons, cover and concealment, route selection, and skill in unobserved day or night movement.

1001. PURPOSE

When an infantry unit is not actively fighting the enemy, it should be actively searching for the enemy, which is the primary purpose of scouting. The unit attempts to keep the enemy off balance while making preparations for further attacks. Physically locating and keeping the enemy off balance are normally accomplished by small units ranging from a two-person scouting party to a squad-size patrol.

Infantrymen are sent out as scouts or as members of a patrol because the commander needs information about the enemy, terrain, and the location of friendly troops. The lives of the entire unit may depend upon the success or failure of a scout or patrol and the accuracy and timeliness of the report. The success of the scout or patrol will depend upon their training, preparation by the commander, and understanding their mission and the commander's requirements.

To wage combat successfully, a commander must have accurate, detailed, and timely information about the enemy, the terrain, and adjacent friendly units. Well-trained scouts and capably led patrols are among the most effective means the commander has for acquiring the information necessary to plan tactical actions and make decisions in execution.

1002. REQUIRED SCOUTING SKILLS

To be effective, a scout must be able to—

- Recognize terrain features.
- Read a map and determine direction.
- Practice and implement the principles of cover and concealment.
- Fully utilize movement and route selection.
- Know the enemy (estimate enemy unit composition and strength).
- Observe and report information accurately.
- Select routes and move through numerous types of terrain.

CHAPTER 2. TERRAIN, MAPS, AND DIRECTION

A scout must understand map symbols, identify elevations from contour interval lines, scale distance on a map, relate natural and man-made features shown on the map to the actual features on the ground, plot a course from one point to another, and locate his current position. To relate a map to the actual terrain and its features, a scout must be able to orient it to the ground using a compass, two points, a watch, and the sun or the stars.

2001. TERRAIN FEATURES

Since the infantry works and fights on the ground, terrain ashore information that scouts gather and report on is of great importance. Hills, valleys, woods, and streams are the forms and growths commonly referred to as natural land features. Artificial or man-made features include houses, bridges, and railroads. Figure 2-1 shows some important terrain features.

2002. THE LENSATIC COMPASS

The best method of finding direction, during both day and night, is with a compass. The lensatic and M2 are the two types of compasses issued to infantry units. Both work on the same principle.

The standard compass for general use in the Marine Corps is the pivot-mounted lensatic compass, so called because azimuths are read through a magnifying lens in the eyepiece. Figure 2-2, on page 2-2, shows the lensatic compass and its nomenclature. The meter graphic scale on the side of the compass is 1:50,000, which is the most used scale in military mapping. The graphic scale is useful in the field as a straightedge, an aid in orienting the map, and a means of reading map grid coordinates. The plastic dial is graduated in both degrees and mils. Numbers on the dial are printed in black. There are luminous markings on the bezel, floating dial, and on both ends of the sighting wire, plus a 3-degree bezel serration and clicking device that permit reading azimuths at night. The compass is carried in a nylon case that may be attached to the cartridge belt.

Magnetic compasses are affected by the presence of iron or magnetic fields. Consequently, the scout should not be within the influence of local magnetic

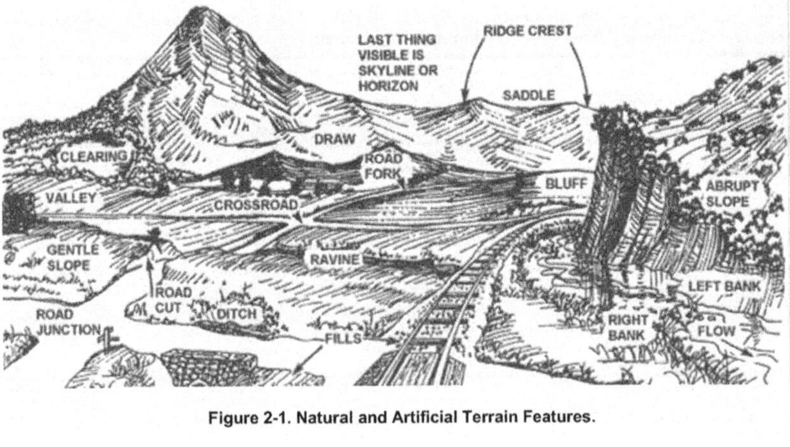

Figure 2-1. Natural and Artificial Terrain Features.

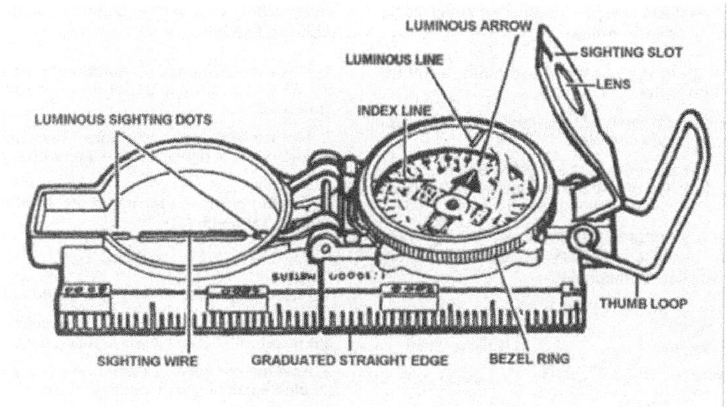

Figure 2-2. Lensatic Compass.

attraction while using a compass to determine direction. The rifle, pistol, and other metal objects must be laid aside when reading the compass. The minimum distances the scout should be from visible masses of iron and electrical fields of magnetism for the compass to provide accurate readings are provided below.

Magnetic Fields	Minimum Distances (meters)
High tension powerlines	60
Field guns	20
Vehicles (wheeled or tracked)	20
Telephone and telegraph wires	10
Barbed wire	10
Machine gun	3
Rifle, pistol	1

Reading an Azimuth

To read an azimuth to any point, the cover of the compass is raised to an angle of 90 degrees in relation to the index face, and the eyepiece is lifted to a 45-degree angle in relation to the bezel, or so the numbers on the dial can be seen. The thumb of either hand is placed in the thumb loop, the index finger extended along the side of the compass case, and the remainder of the hand closed. The closed hand and wrist are grasped with the other hand. The elbows are drawn in close to the body, forming a firm foundation for the compass. The eye is placed to the lens on the eyepiece. The compass is pointed at the object or point to which the azimuth is to be read. A sighting is taken through the sighting slot in the eyepiece, and the point is lined up with the sighting wire in the cover. The compass is held until the dial steadies; then the reading is taken through the lens of the eyepiece. This reading is the magnetic azimuth of the line from the observer to the point.

Reading a Back Azimuth

A back azimuth is the direction opposite the line of sight. If the azimuth is less than 180 degrees, the back azimuth is obtained by adding 180 degrees. If the azimuth is greater than 180 degrees, the back azimuth is obtained by subtracting 180 degrees. Back azimuths are used to determine a return route or to resection to determine a current position.

Circumventing Obstacles

When a scout is traveling on an azimuth and comes upon an obstacle—such as a contaminated area, minefield or swamp—the following steps (sometimes referred to as the *90-degree offset* method) is employed to go around or circumvent the obstacle and resume movement along the original azimuth (see fig. 2-3). The steps are as follows:

1. Move up to the obstacle and make a full 90-degree turn to the right (or left).

2. Walk beyond the obstacle, keeping track of the distance in paces or meters.

3. Stand at the end of the obstacle, face in the original direction of march, and follow that azimuth until the obstacle has been passed.

4. Make a 90-degree turn to the left (or right) and move the distance previously measured to return to the original line of march.

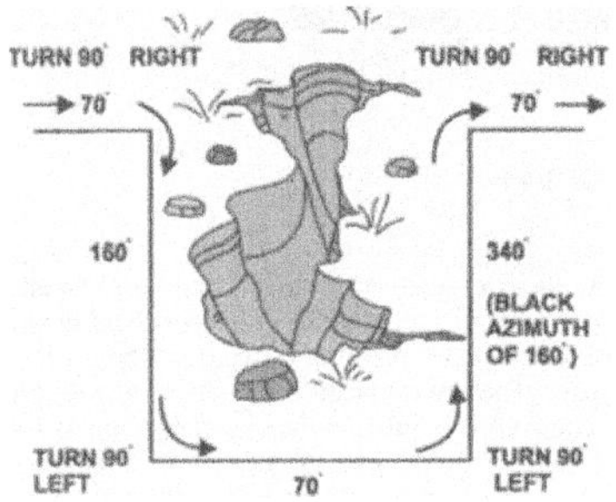

Figure 2-3. Circumventing an Obstacle.

Following an Azimuth During the Day

The eye is placed to the lens in the eyepiece and the compass moved until the desired azimuth reading is visible beneath the fixed index. Without moving the compass, the vision is shifted from the lens through the sighting slot in the eyepiece, and a sighting is taken out beyond the sighting wire in the cover. A prominent terrain feature on this line of sight is selected, the compass closed, and the landmark approached. When the scout reaches the landmark, the procedure is repeated.

Following an Azimuth at Night

It is necessary to prepare and set the compass before departing on a night movement because at night only the luminous parts of the compass can be seen. To prepare the compass for night use, the luminous parts must be fully charged by sunlight or artificial light, such as a flashlight. To set a compass—

1. Move the compass so that the desired azimuth on the dial is directly under the index line on the lower glass.

2. Rotate the upper movable glass so that the luminous line is directly above the north arrow of the dial.

3. Set the compass for marching at night on the specified azimuth.

—OR—

1. Face the general direction of movement.

2. Line up the north arrow and the luminous line on the bezel with the luminous sighting dots.

3. Hold the compass still with one hand and grip the knurled bezel ring with the other hand.

4. Turn the bezel ring the prescribed number of clicks in the proper direction, remembering that each click equals 3 degrees. For example, to set an azimuth of 21 degrees, the bezel ring would be turned seven clicks to the left.

5. Turn the whole compass until the north needle lines up with the luminous line. The compass is then set on the desired azimuth. The azimuth is the line formed by the two luminous sighting dots on the inside of the cover.

To march on a preset azimuth during night movement, open the compass and move it so the north arrow is directly below the luminous line (see fig. 2-4). Move

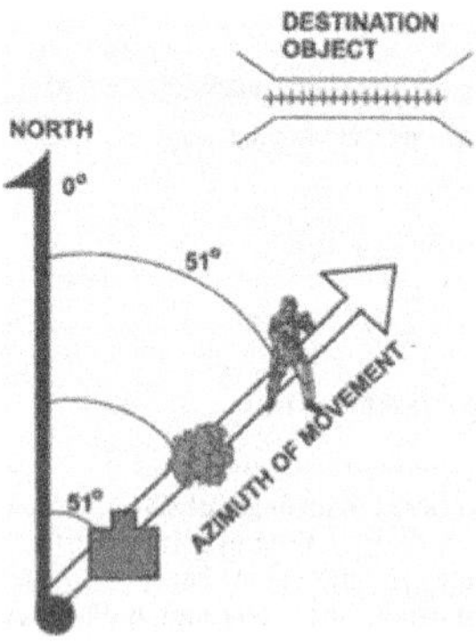

Figure 2-4. Following a Night Azimuth.

in the direction of the line formed by the two luminous sighting dots. It is necessary to refer to the compass more frequently at night than during the day. If stars are visible, find a prominent star along the azimuth of movement to use as a reference point. When the view of the sky is restricted by overcast conditions or vegetation, send a scout forward along the azimuth of movement to the limit of visibility. This scout is guided along the azimuth of movement by a stationary navigator. When the scout reaches the limit of visibility, the navigator moves to the scout's location. This process is repeated until the destination is reached.

A more rapid method for reaching the scout's destination is to equip the navigator with a compass. The navigator can set the compass as explained earlier and the scout proceeds providing security 180 degrees to the front on the specified azimuth, receiving right and left corrections from the navigator while both are on the move. The point scout must stay within visual range of the navigator. If available, a strip of white or luminous tape on the back of the point scout's helmet will assist.

Intersection

Intersection is the location of an unknown point by successively occupying at least two, preferably three known positions and sightings on the unknown point. It is used to locate features not depicted on the map or not readily identifiable. To determine an intersection, perform the following steps (see fig. 2-5):

1. Orient the map using the compass.
2. Locate and mark your position on the map.
3. Measure the magnetic azimuth to the unknown position; then convert to grid azimuth.
4. Draw a line on the map from your position on this grid azimuth.
5. Move to a second known position from which unknown point is visible. Locate this position on the map and again orient the map using the compass. The second unknown position should be a minimum of 30 degrees offset from the first position.
6. Repeat steps 4 and 5.

Figure 2-5. Intersection.

To check accuracy, move to a third position and repeat steps 1 thorough 4. Where the lines cross is the location of the unknown position. Using three lines, a triangle is sometimes formed—called the *triangle of error*—instead of an intersection. If the triangle is large, recheck your work to find the error. Do not assume that the position is at the center of the triangle.

Resection

Resection is the location of the user's unknown position by sighting on two or three known features that are identifiable on the map. To determine a resection, perform the following steps (see fig. 2-6):

1. Orient the map using the compass.
2. Locate two or three known positions on the ground and mark them on the map.
3. Measure the magnetic azimuth to a known position then convert to grid azimuth.

4. Change the grid azimuth to a back azimuth and draw a line on the map from the known position back toward the unknown position.

5. Repeat step 3 and step 4 to determine a second known position.

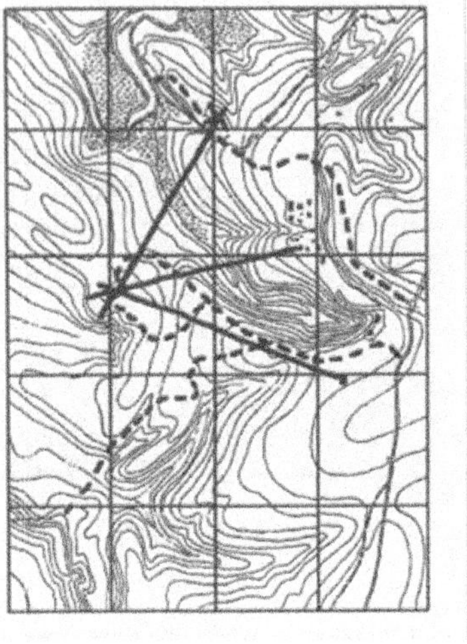

Figure 2-6. Resection.

To check accuracy, repeat the steps above for a third known position. The intersection of the lines is your location. Using three lines, a triangle of error may be formed. If the triangle is large, recheck.

2003. STARS, SUN, AND OTHER FEATURES

In rare cases when a scout is without a compass, the following examples are alternate means to determine direction. When using constellations to determine direction, identify your location's Temperate Zone. A Temperate Zone is the area between the tropics and the polar circles.

At night, the stars provide an excellent means of maintaining a line of march. In the North Temperate Zone (north of the equator), the Big Dipper constellation is one key to determining direction of true north. It is made up of seven fairly bright stars in the shape of a dipper with a long curved handle (see fig. 2-7). The two stars that form the side of the cup farthest from the handle, used as pointers, are situated in the direction of a bright star that is about five times the distance between the two stars of the dipper cup. This bright star is the North Star and is directly over the North Pole. The pointers always designate the North Star, which is the direction of true north.

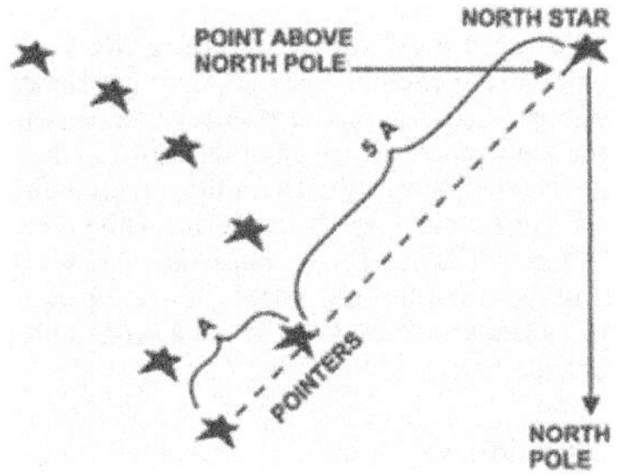

Figure 2-7. Locating the North Pole.

In the Southern Hemisphere, true south is determined in relation to the Southern Cross, a constellation composed of five stars. Two bright pointer stars in the vicinity of the Southern Cross serve as locators to help locate true south (see fig. 2-8). The outer four stars are

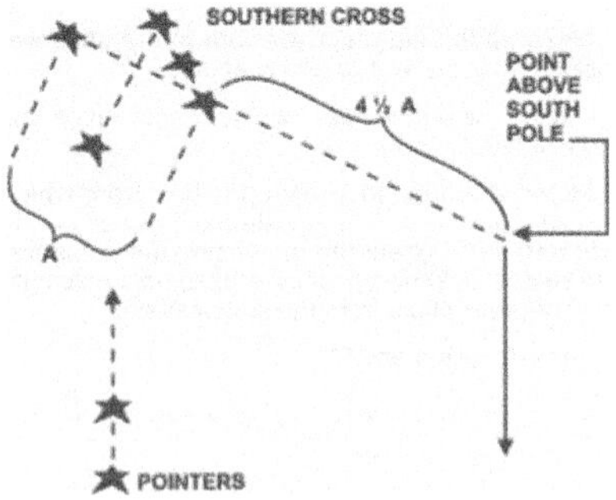

Figure 2-8. Locating the South Pole.

fairly bright and form a cross. This cross is imagined as the frame of a kite. A straight tail, four and one half times as long as the length of the kite itself, is put on the kite using finger widths for a measuring stick. The end of this tail will be close to a position directly over the South Pole. Usually, it will not be possible to see a star in the immediate vicinity, because there is no bright star visible directly above the South Pole.

During daylight hours, a watch and the sun can be used to determine direction within 8 degrees. In the North Temperate Zone, the watch is held horizontally, face up, and the hour hand pointed at the sun (see fig. 2-9). The north-south line and the direction of south can be found midway between the hour hand and the number 12, if the watch is set on standard time. If in daylight savings time, the direction of south is found midway between the hour hand and the number 1.

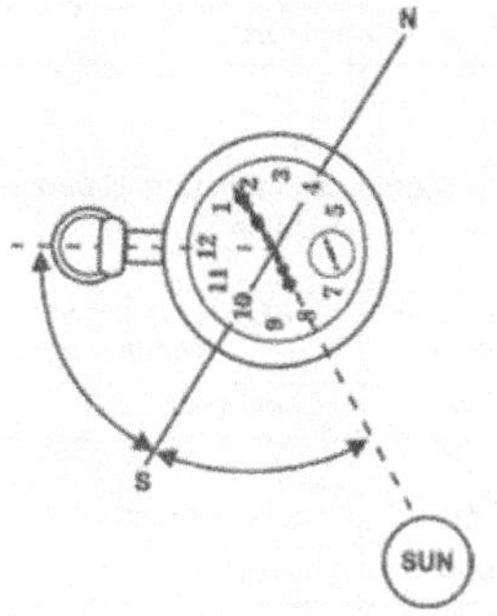

Figure 2-9. Determining Direction by Watch and Sun (North Temperate Zone).

In the South Temperate Zone, if the watch is set on standard time, the number 12 on the watch is pointed at the sun; if the watch is set on daylight savings time, the number 1 is pointed at the sun. North is midway between 12 (or 1) and the hour hand (see fig. 2-10).

When laying in a north-south line, if any doubt exists as to which end of the line is north, remember that the sun is in the east before noon and in the west in the afternoon.

In addition to the sun and stars, other methods a scout without a compass can use to determine direction include determining prevailing wind direction and

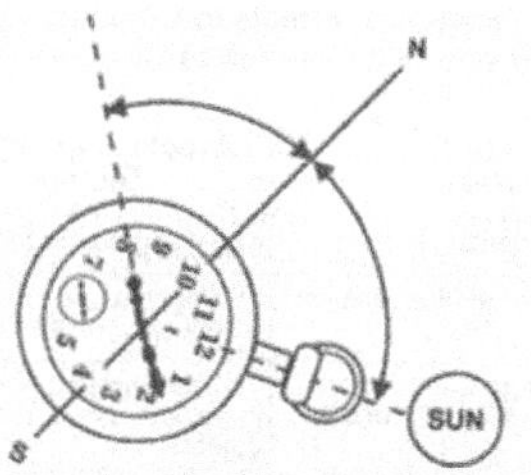

Figure 2-10. Determining Direction by Watch and Sun (South Temperate Zone).

using a mountain for orientation. By previous study of maps and photographs, a scout can keep informed of location and direction by using a distinctive edge of woods, a deep ravine or the direction of a stream's flow. A scout should constantly evaluate and memorize both the immediate terrain and general area for prominent features and landmarks.

2004. RANGE DETERMINATION

Range determination is the method of finding the distance between an observer and an enemy target or an object. By accurate range determination, the members of a given unit can set their sights correctly and place effective fire on enemy targets. The degree of accuracy is dependent on several factors, such as terrain relief, time available, and experience of the observer.

Mental Estimation

A mental distance estimate is made using a known unit of measure. Distance is estimated to the nearest 100 meters by determining the number of known units of measure between the observer's position and a target. For example, a football field, which is 100 yards, can be used as a known unit of measure for determining the distance between an observer's position and a target. For longer distances, progressive estimation may be necessary. To do this, the observer determines the number of units of measure to an intermediate point and doubles the value. The observer should consider the effects in table 2-1 in estimating distances.

Table 2-1. Effects to Consider in Mental Estimation of Distances.

Objects Appear Nearer	Objects Appear More Distant
In bright light.	In poor light or in fog.
In clear air at high altitude.	Only a small part of the object can be seen.
The background is in contrast with the color of the object.	The background is similar in color to that of the object.
The observer is looking down from a height.	The observer is looking over a depression, most of which is visible.
The observer is looking over a depression, most of which is hidden.	The observer is kneeling or sitting, especially on a hot day, when the ground is moist.
The observer is looking down a straight feature such as a road.	
The observer is looking over water, snow, or a uniform surface such as a cultivated field or desert.	

Estimating in Good Visibility

When visibility is good, distances can be estimated by using the appearance of tree trunks, branches, and foliage (as seen by the naked eye) in comparison with map data. Table 2-2 is a guide for wooded terrain. Table 2-3 is a guide for urban environments.

Estimating From a Terrain Study

The Marine should always use terrain/map analysis to assist in estimating distances. When the Marine is looking in a specific direction, the estimation of distance can be enhanced by studying the terrain and comparing it with the map. Particular emphasis should be given to color contrasts of terrain features seen along the observer-target line (OTL). For example, the distance across successive ridge lines or depressions in the distance may be identifiable by only slight changes of color to the eye. Different colors of grass might reveal a hidden terrain feature such as a stream.

Table 2-2. Estimating Distance in Wooded Terrain.

Distance in Meters	Tree Description
1,000	Trunk and main branches are visible. Foliage appears in cluster-like shape. Daylight may be seen through the foliage.
2,000	Trunk visible, main branches distinguishable, foliage appears as smooth surface. Outline of foliage of separate trees distinguishable.
3,000	Lower half of trunk visible. Branches blend with foliage. Foliage blends with adjoining trees.
4,000	Trunk and branches blend with foliage and appears as a continuous cluster, smooth in appearance. Movement of foliage due to wind cannot be deleted.
5,000 and beyond	Whole area covered by trees and appears smooth and dark.

Table 2-3. Estimating Distance in Urban Terrain.

Distance in Meters	Object Identified by the Unaided Eye
1,000	Lone tree trunk
1,500	Individuals and horsemen
3,000	Chimneys on rooftops
4,000	Windows in houses
4,000–5,000	Individual houses in populated area
8,000–9,000	Villages and individual houses
15,000–18,000	Large houses, towers, and steeples

CHAPTER 3. ENEMY ACTIVITY

A commander often acts on information furnished by scouts. Therefore, scouts must aim at absolute accuracy in reporting enemy activity. This chapter discusses estimating enemy strengths, interpreting signs and tracks, and knowing the enemy.

3001. ESTIMATING ENEMY STRENGTHS

If troops cannot be counted, their strength may be estimated by: noting the length of time it takes various types of moving columns to pass given point, the area required of a unit in camp or bivouac, or the front on which they are deployed. When the ground is dry, infantry on the march raise a low, thick cloud of dust, and motor vehicles or mechanized units raise a thick, rapidly moving cloud. Additionally, through practice, a scout may gain information as to the strength and composition of enemy forces by listening to noises and observing lights, fires, and smoke. A scout gains valuable experience in estimating enemy strengths by observing friendly forces in camp, on the march, and deployed. The knowledge scouts gain during field exercises of the appearance and tactical dispositions of squads, platoons, companies, and larger units will be of great assistance in estimating the strength and composition of enemy units observed under various conditions.

3002. INTERPRETING SIGNS AND TRACKS

In addition to estimates made through direct observation, a scout may often be able to estimate size, composition, direction, rate of movement, condition, discipline, state of training, and morale of enemy forces through signs and tracks left behind.

Signs

The examination of vacated enemy positions provides valuable information. The size of a bivouac or defense area ordinarily indicates the number of enemy occupants. Clothing, ration containers, dumps, etc., further indicate the quantity of the departed enemy force. The condition of the bivouac area and amount of material abandoned give an indication of the enemy morale, training, and discipline. A well-policed area indicates good discipline. Rubbish, ration and smoking residue, and nonessential personal items of equipment adrift indicate a lower state of morale, training, and discipline. Stores and material left behind in good condition may indicate a hasty movement or withdrawal. Burned or destroyed materials indicate a deliberate, orderly withdrawal or movement. Letters, insignia, and other articles may reveal the identity of the enemy unit.

In the case of a moving enemy, the distance between periodic halts indicates the rate of march if enemy habits relative to marches and halts are known. Condition of the halt areas indicates the state of morale, training, and discipline.

The physical condition of enemy dead and wounded and their personal equipment and weapons are reported. The general condition and state of maintenance of destroyed or abandoned vehicles should also be reported.

Tracks

A track is a mark left on the ground by the passage of a person or object. Examination of tracks reveals information about the enemy.

Troops

A few tracks overlapping each other on both sides of a road or trail may indicate a patrol in staggered formation. A large number of tracks indicates troops in column formation. A large column will wear a dry road smooth and flat. In damp terrain, a freshly made track will have sharp edges; ordinarily, signs of moisture will disappear in about 15 minutes. A runner's toes are dug into the ground; a walker's footprint is fairly even.

Vehicles

The type of track indicates whether the vehicle is wheeled or tracked. A scout acquires the necessary experience to make the proper determination by observing vehicle tracks during training.

The direction of travel can be determined by the way tracks pass across ruts, by impressions on the edges of holes in the ground, how water is splashed from puddles, or by the way grass, twigs, and branches are bent; for example—

- A vehicle (wheeled or tracked) entering a rut pushes dirt into the rut and leaves an indentation on the exit side of the rut.
- A wheel going over holes in the ground leaves a deeper impression on the edge toward the direction of travel.
- The side of a puddle with the greater splash indicates the direction of travel of the vehicle.
- When traveling cross-country, the direction in which grass is bent and/or twigs, branches, and bushes are broken indicates direction of travel.

A general rate of speed can be estimated by the amount of water or mud splattered. A fast-moving vehicle will throw larger amounts of water or mud a greater distance to the front and sides than a slow-moving vehicle, and it leaves a deeper impression on the exit edges of holes. The faster the travel, the deeper the impression.

3003. KNOWING THE ENEMY

A scout should learn as much as possible about enemy psychology, habits, organization, and tactics. The more knowledge gained about the enemy, particularly the enemy's normal security measures, the better the scout's chances are to observe and obtain accurate information with minimum risk to the mission's success. Scouts gain much of this knowledge through experience, but they also gain a great deal of their preliminary information, particularly that pertaining to enemy organization and tactics, during training and may be updated by unit commanders and intelligence officers.

CHAPTER 4. DAYLIGHT SCOUTING

A scout must be able to operate in all types of terrain and under all conditions of visibility. He must be thoroughly familiar with the principles for using cover and concealment, camouflage, individual movement, and route selection, both to and from the objective.

4001. COVER AND CONCEALMENT

Cover is protection from the fire of hostile weapons. Concealment is protection from observation or surveillance from hostile air and ground observation, but not from hostile fire. Both cover and concealment are divided into two main categories: natural and artificial. Natural cover includes small hills, ditches, rocks or vegetation. Fighting holes, bunkers, and brick walls are examples of artificial cover. Some features, such as buildings, provide both cover and concealment. In deciding whether to seek cover or concealment, a scout must make the best choice to complete the mission (see fig. 4-1).

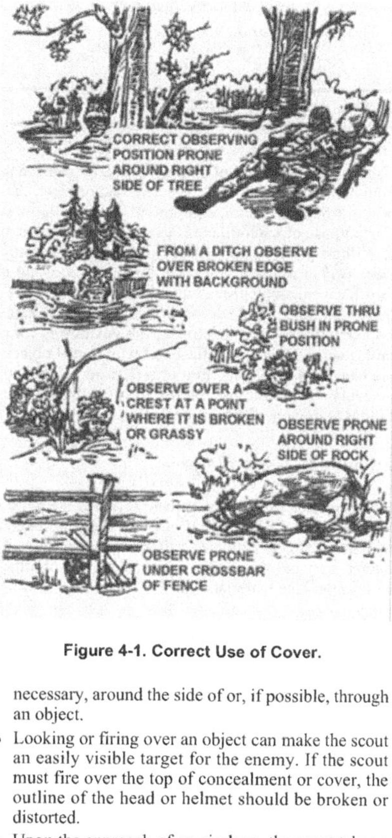

Figure 4-1. Correct Use of Cover.

Concealment Principles

Concealment principles are as follows:

- Remain motionless while observing. Anything in motion attracts the eye.
- Use all available concealment.
- Observe from the prone position (it offers a low silhouette and makes detection by the enemy difficult.
- Expose nothing that reflects light.
- Blend with the background because contrasting colors are noticeable.
- Remain in the shade because moving shadows attract attention.
- Distort or change the regular outline of objects. Most military objects have distinctive shapes that make obvious shadows and silhouettes.
- Avoid the skyline. Figures on the skyline can be seen from great distances and are easily identified by their outlines.

Concealment Techniques

Concealment techniques are as follows:

- When observing, the scout looks around an object's side (unless it is transparent) and prepares to fire, if necessary, around the side of or, if possible, through an object.
- Looking or firing over an object can make the scout an easily visible target for the enemy. If the scout must fire over the top of concealment or cover, the outline of the head or helmet should be broken or distorted.
- Upon the approach of an airplane, the scout takes a prone position, turns face-down, and remains motionless. If surprised by an airplane, the scout remains in place and does not look up.

- The scout covers exposed body parts such as the face, back of the neck, and hands with grease paint, mud or other materials to reduce sun reflection.
- Camouflage for equipment can be improvised from garnishing or sandbags to prevent sun reflection.
- In snowy terrain, white overgarments are worn.
- The helmet cover outline should be distorted.

4002. CAMOUFLAGE

Camouflage is the use of concealment and disguise to minimize the possibility of detection and/or identification of troops, material, equipment, and installations. The purpose of camouflage is to provide concealment of military objects from enemy observation. Camouflage is also used to conceal an object by making it look like something else. A scout's mission usually requires individual and equipment camouflage. If natural camouflage is not adequate, the position is camouflaged. In using camouflage, remember that objects are identified by their form (outline), shadow, texture, and color. The principal purpose of camouflage in the field is to prevent direct observation and recognition.

Individual Camouflage

Successful individual camouflage involves the ability to recognize and take advantage of all forms of natural and artificial concealment available (vegetation, soil, debris, etc.) and knowledge of the proper use of artificial camouflage materials.

Aids to Individual Camouflage

A scout must recognize the terrain's dominant color and pattern and must change the appearance of clothing and equipment accordingly in order to blend and not contrast with the terrain (see fig. 4-2).

The helmet is camouflaged by breaking up its shape, smooth surface, and shadow. Use of a helmet cover works best. In the absence of a helmet cover, mud can be irregularly blotched on the helmet to disguise its form and dull the surface. A helmet cover may be improvised from irregularly colored cloth or burlap to blend with the background. Foliage can be draped to prevent the visor of the helmet from casting a dark shadow across the face. Foliage should not stick up like plumes because any head movement will give away the position.

A small, thin bush in the shadow of a large bush makes a good observation point. Lone trees, rocks, fence corners, and outstanding landmarks are easily picked up by the enemy as obvious observation posts.

If camouflage clothing is not available, other available clothing can be attached in irregular splotches of appropriate colors.

Exposed skin reflects light and attracts the enemy's attention. Even very dark skin will reflect light because of its natural oil. The buddy system is recom-mended when applying camouflage. Standard Marine Corps issue camouflage face paint sticks are two toned:

- Loam and light green for light-skinned troops, in all but snow regions.
- Sand and light green for dark-skinned troops.
- Loam and white for troops in snow-covered terrain.

Shiny areas (forehead, cheekbones, nose, and chin) are painted with a dark color. Shadow areas (around the eyes, under the nose, and under the chin) are painted

Figure 4-2. Avoid Contrasting Backgrounds.

with a light color. Skin that is exposed on the back of the neck and hands is painted with a two-color combination in an irregular pattern (see fig. 4-3).

When standard issue face paint sticks are not available, burnt cork, charcoal or lamp black can be used to tone down exposed areas of skin.

Mud is used only in an emergency because it changes color as it dries and may peel off, leaving the skin exposed. Since mud may contain harmful bacteria, mud should be washed off as soon as possible.

Any equipment that reflects light should be covered with a nonreflective material that aids in the concealment of the weapon (for example, black electrical tape or mud). The straight line of the rifle or other infantry weapons may be very conspicuous to an enemy observer. The barrel and hand guard should be wrapped with strips of contrasting colored cloth or tape to break the regular outline. Mud or dirt dulls the reflecting surface of the stock, barrel, and bayonet where coloring has been worn. Lamp black may also be used on metal parts. The function of the weapon must not impaired.

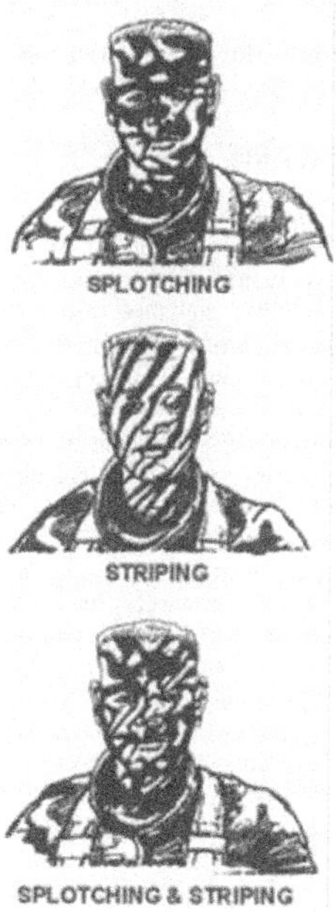

Figure 4-3. Face Camouflage.

If time, material, and surroundings permit, a ghillie suit should be constructed. (Refer to MCWP 3-15.3, *Scout Sniping*.)

Aids to Camouflage a Position

To successfully camouflage a position, the scout must remember to—

- Camouflage the position as soon as it is occupied.
- Avoid using too much material for camouflage. Even though natural materials are used, too much may make the object and its shadow stand out from its surroundings, thus attracting the attention of a hostile observer.
- Inspect completed camouflage work from the enemy's point of view to check effectiveness.

Continuous Camouflage

Camouflage around and on the scout's position must be maintained in a fresh condition as wilted and dead foliage can give the position away. If the mission dictates that the position should be occupied for longer periods, wilted foliage should be replaced during periods of reduced visibility.

4003. INDIVIDUAL MOVEMENT

Principles

The principles of individual movement are as follows:

- Scouts move from one concealed position to another. When not changing positions, they remain motionless.
- The scout's head is lifted slowly but steadily, without abrupt movements, to search for a new position.
- Scouts select the next stopping place before moving and ensures it is not contained by the enemy.
- Scouts change position on the run: springs up, runs with the body bent low, zigzags, quickly drops to the ground slightly to the right or left of the objective, then rolls or crawls to the desired position.

(Remember the phrase, "I'm up—he sees me—I'm down.")

Rushing

When starting from the prone position—

- Raise the head slowly and steadily and select a new position.
- Lower the head slowly, draw arms inward, cock right leg forward, and prepare to rush.
- Use one movement to raise the body by straightening both arms.
- Spring to your feet, stepping off with the left foot.
- Bend forward as low as possible when running. Never advance directly to the next position; always zigzag.

When hitting the deck—

- Stop.
- Plant both feet in place.
- Drop quickly to the knees and slide the hand to the heel of the rifle.
- Fall forward, breaking your fall with the butt of the rifle. (To confuse the enemy, roll over after hitting the deck and roll into firing position with feet, knees, and stomach flat on the ground.)
- Keep head down if you do not intend to fire.

When rolling over—

- Hit the deck and assume the prone position.
- Bring the rifle in close to the body, placing the rifle butt in the crotch.
- Roll over swiftly to confuse any enemy observers as to final intended location. Never reappear at the same place you went down.

Low Crawl

The low crawl is used when—

- Cover and concealment are scarce.
- The enemy has good observation over the area in which the scout is moving.
- Speed is not essential.

To perform the low crawl, keep the body as flat as possible against the ground. Grasp the rifle sling at the upper sling swivel. Let the balance of the rifle rest on the forearm and let the butt of the rifle drag on the ground. Keep the muzzle off the ground.

To start forward, push arms forward and pull right leg forward. To move forward, pull with arms and push with right leg. Change the pushing leg frequently to avoid fatigue.

High Crawl

The high crawl is used when—

- Cover and/or concealment are available.
- Poor visibility reduces enemy observation
- Greater speed of movement is required.

To perform the high crawl, keep body off the ground. Rest weight on forearms and lower legs. Cradle rifle in arms, keeping the muzzle off the ground. Keep knees well behind the buttocks to stay low.

Move forward, alternately advancing right forearm and left knee; then left forearm and right knee.

Movement Aids

Aids to movement include—

- Carrying only necessities. Additional weight causes premature fatigue and impedes free movement.
- Not disturbing birds or animals whose flight would betray your presence. If birds or animals are alerted, remain motionless under cover for a few minutes, as the enemy's attention may also be attracted.
- Moving during an incident that diverts attention, such as an airplane flight, a distant disturbance or sudden bursts of fire.
- Fog, smoke, or even light haze offer concealment for movement; however, the enemy may have thermoimagery and night vision devices. Therefore, darkness and smoke cannot be used as easily.
- Following a stream or road by staying as far away from them as possible while still keeping them in sight. Keep close to the dune line when moving along a beach.
- When moving through tall grass or similar growth, move when the wind blows, changing direction frequently. A straight route will be more readily noticed.
- Whenever possible, avoid areas of soft ground so as not to leave tracks.

- When crossing a road or water obstacle, choose crossing sites where the enemy's observation is restricted (an area in shadows or near a bend) and cross rapidly.

4004. ROUTE SELECTION

Prior to Movement

A scout and the immediate commander conduct a map reconnaissance before starting on a mission. This assists them in selecting the route according to available cover and concealment and any indicated enemy activity.

Prior to and during the course of the mission, move to an observation point to visually reconnoiter the terrain for movement and select the tentative route. It may be necessary to make wide detours around open spaces or those containing enemy patrols or other enemy activity.

Carefully study the country to be traversed and pay close attention to the general features, streams, ridges to be crossed, and their relation to the general direction to be taken (see fig. 4-4).

Make notes of terrain features and landmarks along the proposed route and rely on notes for guidance (see fig. 4-5 on page 4-6). Additionally, determine the compass direction and readings for each change of direction at the start. Finally, learn the location of unit boundaries and observation/listening posts as well as general location of other friendly or scouting parties. Be sure to avoid man-made and natural obstacles as they will slow progress and overall success of the mission. If possible, use the local populous as a source of intelligence. When returning to friendly lines, avoid using the same route.

En Route

En route, the actual advance will be a series of movements from one observation point to the next. The distance and route will depend on cover and terrain. Assess the cover, terrain, and any enemy or civilian activity to determine whether or not to modify the approach or return routes. Unless the mission requires it, avoid danger areas (for example, houses, villages, potential assembly or bivouac areas, roads, and streams) that may give away your position by being

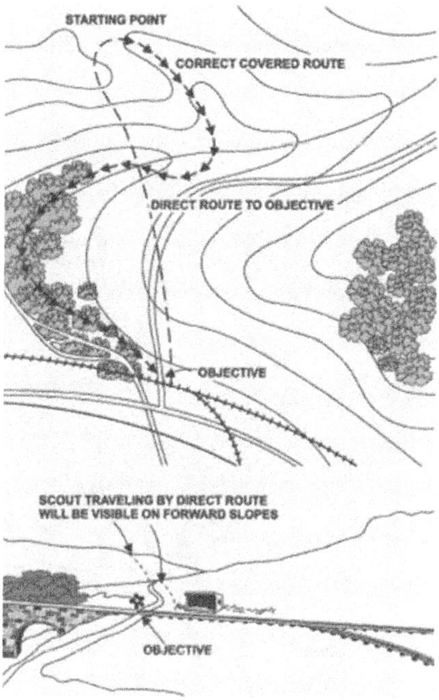

Figure 4-4. Choosing a Concealed Route of Advance from a Map.

observed by the enemy. When required to reconnoiter danger areas, choose a covered approach and return, and make entry or passage as quietly and quickly as possible. If part of a larger effort, the approach and return should be covered by observation and fires of the other members of the scouting party or patrol.

Stream Crossings

When the crossing does not appear to be held by the enemy, advance upon it rapidly. If there are two or more scouts, one crosses while the other(s) provide protection. Note the length, width, depth, and approaches to a crossing. Observe the condition of the road or trail that crosses the stream, and report on the suitability of the crossing for use by tracked and wheeled vehicles. If the crossing is under observation by enemy, seek another crossing site or dash across to avoid detection.

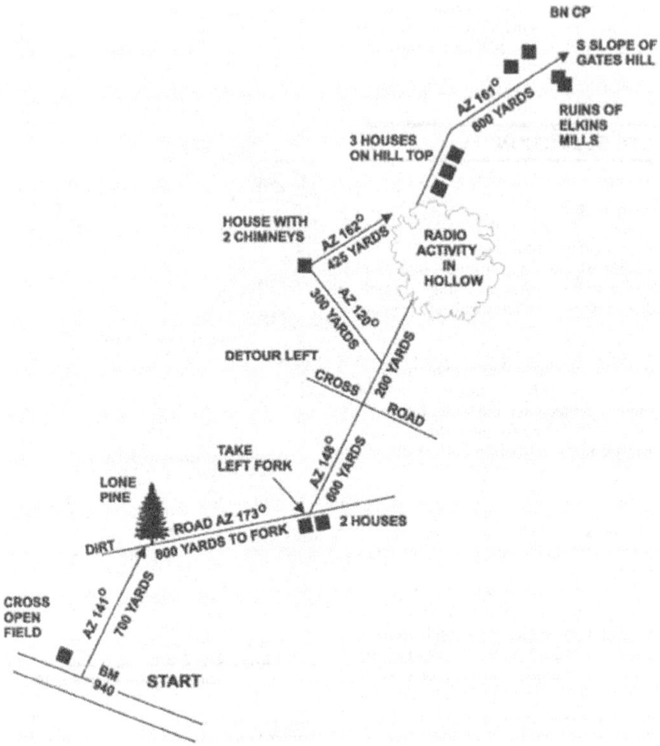

Figure 4-5. Proposed Route Sketch.

CHAPTER 5. NIGHT SCOUTING

Night scouting presents many of the same problems encountered in day operations-such as cover, concealment, movement, and camouflage-as well as additional considerations. Knowledge of human eye construction and operation will enable maximum advantage under night conditions or poor visibility.

5001. NIGHT VISION

Certain parts of the eye correspond to parts of a simple camera (see fig. 5-1). The lens focuses light entering the eye similar to a camera lens. The iris (colored part of eye) corresponds to the diaphragm of a camera, opening and closing to regulate the amount of light entering the eye through the pupil. The retina corresponds to camera film. Light rays strike the retina, form an image, and cause an impression to be transmitted to the brain through the optic nerve. In a camera, the image is formed and fixed on film.

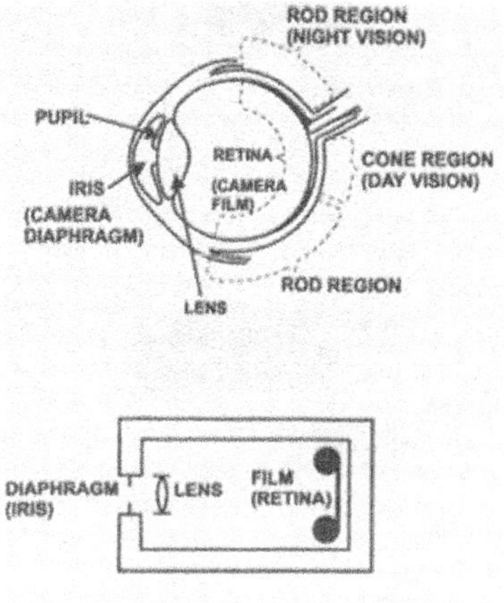

Figure 5-1. The Eye is Like a Camera.

The retina is composed of cone cells and rod cells, so-called because of their shapes. Cone cells distinguish color, shape, and sharp contrast. Because they are activated by light conditions, they are blind during periods of low illumination. Rod cells produce a chemical substance called visual purple that makes them active in darkness, low illumination or night conditions. Rod vision distinguishes black, white, shades of gray, and general outlines.

Principles

To effectively "see" at night, the principles of night vision dark adaptation, off-center vision, and scanning are applied.

Dark Adaptation

Allowing the eyes to become accustomed to low levels of illumination is called dark adaptation. It takes the rod cells about 30 minutes to produce enough visual purple to activate them and enable the eye to distinguish objects in dim light. This may also be accomplished by staying in a red-lighted area, or by wearing red goggles for 20 minutes, followed by 10 minutes in darkness (which allows the pupils to open wide). This method saves valuable time by allowing Marines to be in a lighted area to receive orders, check equipment, or perform some other function before moving into darkness.

Off-Center Vision

The technique of focusing on an object without looking directly at it is called off-center vision. When looking directly at an object, the image is formed on the cone region, which is not sensitive at night (see fig. 5-2 on page 5-2). When looking slightly to the left, right, above or below an object, the image is formed on the area of the retina containing rod cells, which are sensitive in darkness. The most sensitive area varies in individuals, but is usually found by looking 6 to 10 degrees away from an object; in effect, out of the corner of the eye (see fig. 5-3 on page 5-2).

Scanning

Off-center vision used to observe an area or an object is called scanning. When using rod vision, the visual purple in the rod cells bleaches or blacks out in 4 to 10 seconds and the object observed disappears. As the visual purple in the rod cells in one area bleaches out, the eyes must slightly shift to use fresh rod cells. Eyes

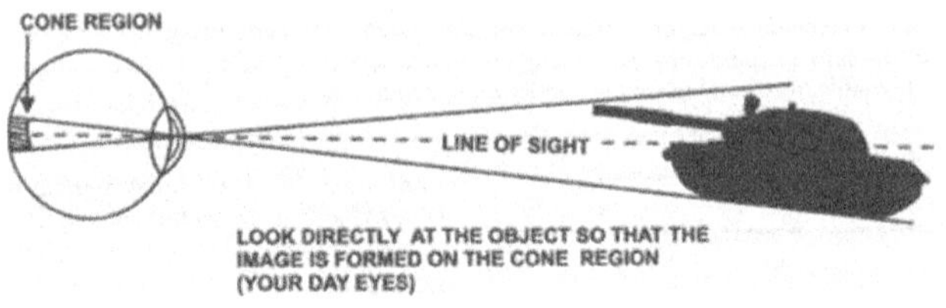

Figure 5-2. Day Vision.

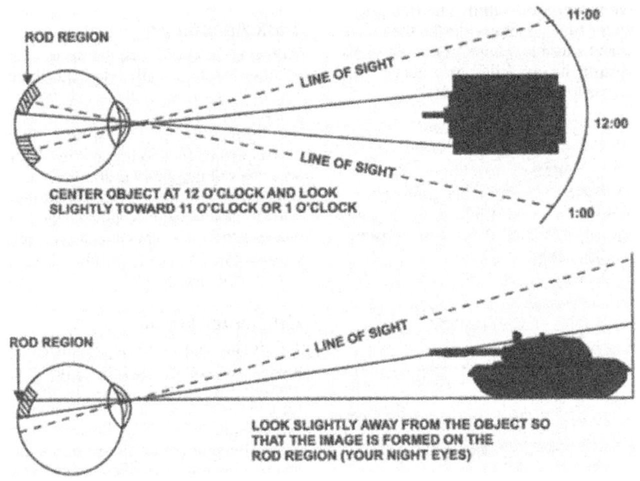

Figure 5-3. Night Vision.

should be moved in short, abrupt, irregular movements over and around the target (see fig. 5-4).

Preserving Night Vision

Night vision is quickly destroyed if bright light is allowed to enter the eye. When entering a lighted area or when observing in a temporarily lighted area (illumination, flares), one eye should be closed and covered to preserve its night vision. When the light goes off, fades or the lighted area is exited, the night vision retained by the protected eye enables it to see until the other eye adapts to the darkness. Red light helps preserve night vision, but like white light, it can be observed at long distances.

Factors that decrease night visual acuity include fatigue, lack of oxygen, long exposure to sunlight, alcohol, nicotine (within the past 48 hours), and age. When night vision has been attained, straining will not improve effectiveness; however, practice identifying objects at night will improve perception.

5002. APPEARANCE OF OBJECTS

Darkness not only makes objects difficult to see but also changes their appearance, distorts size, and blots out details. A tree visible against the night sky appears smaller than in the daytime because the twigs at the end of branches cannot be seen. A scout must train to identify objects by block outlines at night and cannot rely on details visible in daylight. Binoculars enlarge objects or parts of objects otherwise too small to be seen and help identify objects already spotted. Night observation devices increase night visibility and should be used whenever possible.

5003. SOUNDS

At night, sounds become very important. By listening, a scout gains information about the enemy and by exercising care, keeps information from the enemy. A scout stops frequently to listen. Scouts must listen for long periods in perfect silence. Hearing is amplified with the mouth open; removing the helmet will reduce sound distortion. Sounds are transmitted a greater distance in wet weather and at night than in dry weather and in the daytime. By holding the ear close to the ground sounds of people walking and vehicles moving can be heard. Sound travels approximately 370 meters a second. When a flash from a fired weapon is observed, the range to the weapons can be easily estimated by counting the time interval between the flash and hearing the report. For example, counting to three (one thousand one, one thousand two, one thousand three), indicates the distance is 1,110 meters. The cadence is determined by actual practice at known ranges.

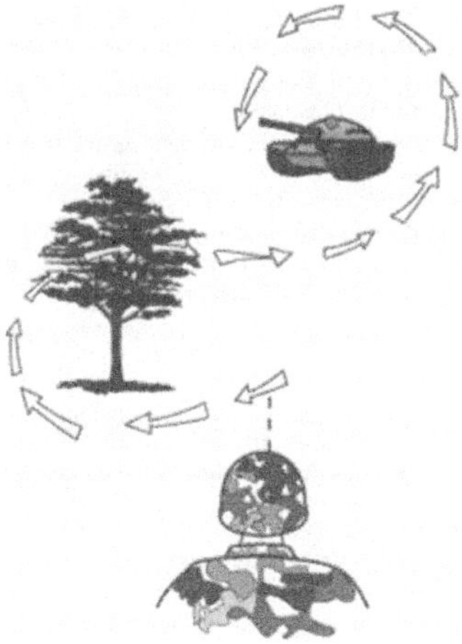

Figure 5-4. Scanning

5004. SMELLS AND TOUCH

A scout's sense of smell can warn of enemy fires, cooking, motor parks, gasoline and diesel engines, and bodies of water. A scout must feel and recognize objects in the dark, and adjust and operate equipment quietly by sense of touch.

5005. CLOTHING AND WEAPONS

All loose clothing must be secured (string or tape can be used) to prevent snagging on barbed wire, brambles, and brush. Helmet covers are worn to muffle sounds made by low branches.

The belt buckle should be turned around to the side in order to move in a prone position without scraping the buckle against stones or hard surfaces. Identification tags can be taped together to prevent rattling. Hands, face, and neck can be blackened so skin does not reflect light or appear as white spots in the darkness. (Refer to para. 4002.)

Scouts are normally armed with rifles. Rifle slings should be taped to prevent rattling. All weapons parts should be checked for glare elimination measures.

5006. CONCEALMENT

Although total darkness provides concealment, scouts must observe the same principles of concealment during moonlight conditions as in the daytime. Scouts should assume enemy employment of night observation devices and observe the principles of night movement such that presence will not be disclosed by noise when close to the enemy.

5007. AIDS TO NIGHT SCOUTING

Aids to night scouting include the following:

- Carry out scouting missions close to or within hostile positions on dark or rainy nights.
- Stifle a threatening sneeze by pressing fingers upward against the nostrils.
- Stifle a threatening cough by applying slight pressure with the finger on the Adam's apple.
- Stop a ringing sound that interferes with hearing by yawning.
- Speak softly rather than whispering if voice communication is necessary.
- Move boldly and rapidly when taking advantage of any sound-such as shelling, rustling wind or distant firing-to push forward if firing is taking place.
- Avoid shell craters and depressions in damp weather conditions if the enemy has employed chemical munitions.
- Move the eyes constantly; concentrating on one object too long will strain them.
- Take notice of the enemy's use of flares. When the enemy employs flares, few enemy patrols are apt to be out; when flares are not employed, the enemy's patrols are likely to be numerous.
- Drop to a prone position upon hearing a flare being fired and before it illuminates. Remain motionless while it is burning. If open or moving when a flare bursts in the air, freeze or drop quickly in the split second after the flare illuminates while the enemy is blinded. You are an easy target for the enemy if the flare bursts in the air or on the ground behind you. Never look at a flare. If you activate a trip flare, drop to the ground and crawl away from the illuminated area.
- Consider all patrols or individuals encountered as hostile until proven friendly. If encountering someone, crouch low, silhouetting the approaching person against the sky. At the same time, make yourself an indistinct target in case the person encountered is an enemy.
- Return fire only to avoid capture if fired on when close to enemy positions.

5008. AIDS TO NIGHT MOVEMENT

- Aids to night movement include the following:
- Move silently.
- Advance in stealthy legs. Each leg should follow some terrain feature that serves as a guide. When there are no terrain features to serve as guides, move in a straight or nearly straight line from one defined point to another, or maintain direction by using a compass.
- Avoid running, except in an emergency.
- Take advantage of sounds that may distract the enemy.
- Fall silently without making an outcry.

Walking

When walking at night—

- Place the heel down first. Balance the weight of the body on the rear foot until a secure spot is found.
- Lift the forward foot high to clear any stiff grass, brush, or other obstruction.
- Continue to balance body weight on the rear foot, lower the forward foot gently, toe first, to explore the ground for objects that might make noise. Step over fallen logs and branches, not on them.
- Lower the heel of the forward foot slowly; gradually transferring body weight to that foot.

Creeping

The low crawl and high crawl are not suitable at night when very near the enemy because an easily heard shuffling noise results. Creeping is the recommended method of movement:

- Creep at night on the hands and knees.
- Use your hands to feel for twigs, leaves or other substances that might make a noise. Clear a spot to place your knee. Keeping your hand at that spot,

bring your knee forward until it meets your hand. Then place your knee on the ground and repeat the action with the other hand and knee.

- Lay the rifle on the ground at your side and clear an area for it. Lift the rifle up and move it forward. Movement is slow and tedious, since it must be done silently.

Hitting the Deck at Night (Right-Handed Shooter)

To safely hit the deck at night from the standing position—

- Advance your left leg, place the butt of the rifle in your right armpit with the hand remaining on the pistol grip, and grasp it with the right hand at the balance.
- Quietly drop down on the right knee and left hand.

- Move the left leg carefully to the rear, and then move the right leg to the rear.

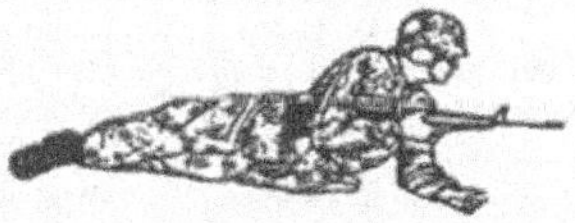

- Lie flat on the ground, or take up a firing position if necessary.

Wire Obstacles

A mission often requires a scout to pass through and work behind enemy positions. To accomplish this, the scout must be able to quietly pass through enemy wire obstacles and cross trenches. Cutting a gap in wire is time-consuming. If possible, walk over the low bands of enemy wire and crawl under the high bands (see fig. 5-5). Avoid movement along wire barriers, as enemy

Figure 5-5. Crossing Wire Silently at Night.

covering fires are generally planned parallel to them to take advantage of canalization and enfilade fire.

To step over low wire at night, crouch low to view the strands against the sky. Grasp the top strand with one hand; with the other hand, reach forward and feel for a clear spot for foot placement without stepping on other strands or any object apt to make a noise. Raise the body up, still grasping the top strand of wire. To avoid catching the foot in another strand, lift the foot up and over, passing it close to the hand grasping the wire.

If a high wire obstacle is encountered at night and wire cutters are not available, pass under the wire with your back on the ground. Grasp the lowest strands in your hands and hold them clear of the body while you slide under them.

When cutting wire and working solo, cut a wire near a post (see fig. 5-6), then dispose of all but one loose end. Grasp the wire close to a post and cut between your hand and the post, muffling the sound and keeping the loose wire in your grasp. When cutting wire in tandem, one firmly holds the wire with the hands positioned close to the cutters, in order to muffle the sound and prevent the loose ends from flying back, while the other one cuts. In both instances, the loose ends of the wire are bent back to form a passage.

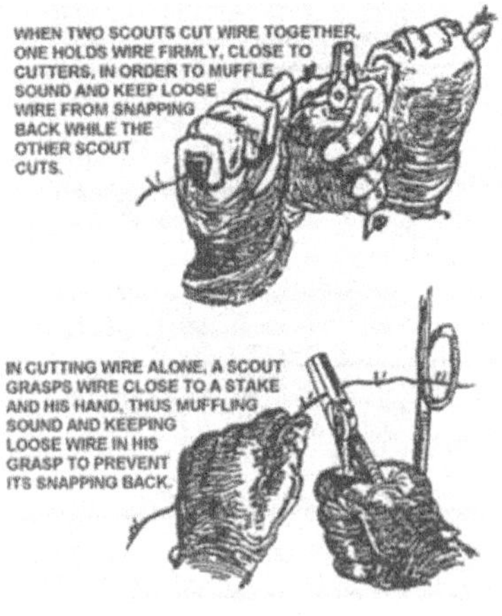

Figure 5-6. Cutting Wire Silently at Night.

Wrap a sandbag around the wire cutters and wire to deaden the sound.

Do not cut a complete gap in the wire; cut only the bottom wire(s). Leave the top wire(s) intact to lessen the chance of discovery by the enemy.

Crossing Trenches

Before approaching a trench, wait outside the trench for awhile and listen. Do not enter or cross a trench near its junction with a communication trench. Crawl silently up to the edge of the trench and look into it. Remove all loose dirt and rocks from the edge. If it is a narrow trench, spring up and jump across, sinking quietly to the ground on the other side and remaining there a moment to listen before proceeding. If the trench is wide, climb silently and slowly down into it and out the other side, using the revetment for support (see fig. 5-7). Do not enter enemy trenches unless it is absolutely necessary in order to accomplish the mission. Ordinarily, work is better accomplished from outside the trench. Sentries usually pay more attention to sounds in front of them; therefore, if it is necessary to enter a trench, cross it first at the place where enemy observation is restricted, then approach from the rear.

5009. LOCATING AND PLOTTING THE ENEMY AT NIGHT

For night work, a scout must understand the use of a lensatic compass. (Refer to para. 2002.) Using the lensatic compass, the scout can guide the platoon into position, locate adjoining elements of the command, keep direction when on patrol, determine the location of gaps in the enemy wire and the position of enemy out guards.

Locating Gaps in Enemy Wire

When searching for gaps in enemy wire, at least two lensatic compasses are needed: one to register the gap in the wire and the other for navigation. When a gap in enemy wire is located, lie outside the gap, keeping a distance of 10 meters from the barbed wire. Sight with the lensatic compass on a prominent point on the skyline in line with the gap. In selecting the prominent point in the skyline, pick one that appears on the map (i.e., hill mass, house, road junction). If the only prominent point available is one not identifiable on a map (i.e., a tree, destroyed vehicle, enemy position),

Figure 5-7. Crossing Trenches Silently at Night.

use it. The exact location of this point can be fixed the next day by visual reconnaissance of the area from an observation point. When the needle rests—

- Clamp it in place by lowering the eyepiece to the closed position.
- Rotate the movable bezel ring until the luminous line is directly over the north end of the needle. The azimuth of the gap from the prominent point is now registered.

The compass is carried back without further adjustment. The azimuth setting can be recorded later on a map (see fig. 5-8).

Locating Enemy Out Guards

At night, locate enemy guards by their sounds and failure to maintain light discipline. As sounds of the enemy are heard, and/or observations of the enemy made, shoot an azimuth with one compass. When the needle comes to rest, clamp the compass by lowering the eyepiece to the closed position.

Figure 5-8. Locating a Gap in Enemy Wire at Night.

Note the time and nature of each sound, the estimated distance, and which compass was used to fix the location. Plot this data on a map then wait until the debrief to turn in the notes and compasses. See figure 5-9.

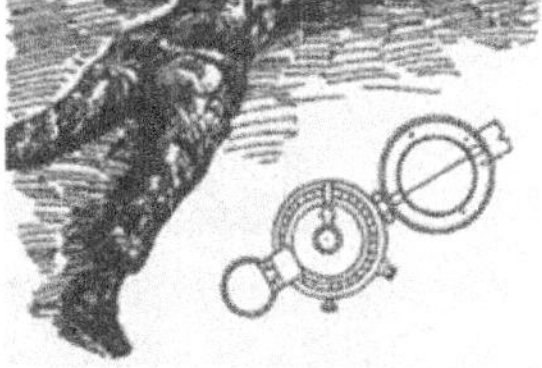

Figure 5-9. Locating Enemy Outguards at Night.

50010. ROUTES OF MOVEMENT

Prior to beginning a night missions, a scout studies the ground in detail from an observation point, air photos, and a map during daylight. The route of advance should be below the skyline. Avoid becoming a silhouette (see fig. 5-10).

Unless the moon is bright, avoid, if possible, passing through woods, ditches, ravines, and brush, because noises of movement may lead to discovery. If the enemy is known to have night observation device capability, avoiding these kinds of terrain may not be possible. To avoid enemy ambushes, return by a different route from the advance and change routes on successive nights.

Figure 5-10. Difference Between Correct Routes by Day or Night.

Chapter 6. Observing and Reporting

When sent out on a mission, a scout's duties are to observe and report. the first five chapters covered scout protection measures and movement in enemy territory. This chapter discusses performing the assigned mission and reporting the mission after its completion.

6001. OBSERVATION POSTS

Positions

The following guidelines apply to observation posts. See figure 6-1.

- When selecting observation posts, scouts should chose the least prominent position.
- Scouts may occupy one or more observation posts.
- An observation post should not be manned for more than 24 hours.
- A selected observation post should be observed for 10 to 15 minutes to ensure it is not occupied.
- Scouts move to the chosen observation post by a concealed route.
- If the post is located on a hill, crawl to a position where the skyline is broken.
- If a tree is used, the position should have a background so as not to be silhouetted against the sky while climbing or observing
- When leaving the observation post, a different route from that of the approach should be used.
- If a radio is used, its antenna should be located to provide clear communication to the controlling commander but masked from enemy observation and direction-finding equipment. Upon departure, scouts should remove the antennae from the observation post so as not to give away the position.

Observing

Using all senses available, be particularly alert for movement, objects, sounds, and smells inappropriate to the surroundings. While observing, avoid all unnecessary movement. If observing from a building, keep back from doors and windows.

In daylight, look first at the ground nearest you. Begin observing close to your post and search a narrow strip 50 meters or less deep, going from right to left parallel to your front; then search from left to right a second and similar strip farther away but overlapping the first. Continue to observe until the entire field of view has been searched (see fig. 6-2 on page 6-2).

At night, use a night observation device. If one is not available, search the horizon with short, jerky movements, and short pauses. Look a little to one side of an object and then to the other. Lower the head close to the ground to view the object more clearly. Use low-powered field glasses to increase sight range.

Figure 6-1. Method of Approaching an Observation Post.

Figure 6-2. Method of Searching Ground.

6002. REPORTING

It is imperative that the scout accurately and completely report who, where, when, and what was observed to the proper authority upon mission completion. In most cases, a scout will not be equipped with a radio.

Verbal Reports

Verbal reports should be made when writing is impractical, when the information is not complicated or when the enemy is likely to intercept a messenger carrying a written message.

Written Messages

Written messages, preferred to verbal reports, are recorded in message book blanks issued for that purpose (see fig. 6-3) and delivered to a higher authority as soon as possible. The NATO spot report (SPOTREP) should be reviewed as the written message is a NATO format.

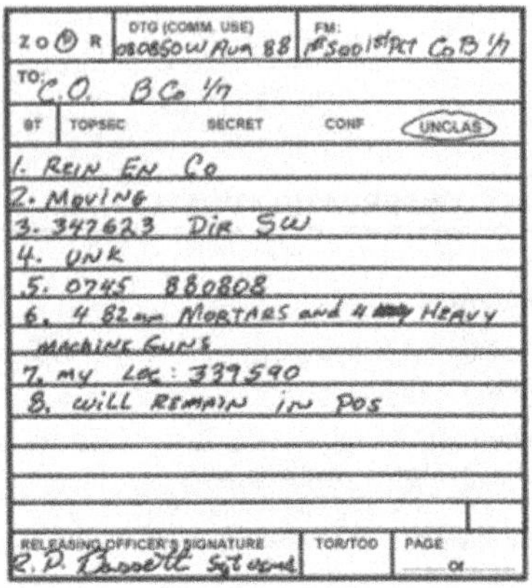

Figure 6-3. Sample Field Message.

The message body is brief, accurate, and clear; facts and opinions are distinguished. If secondhand information is reported, its source is included. Reports include all information of value, first about the enemy, and then about the originator. Information about the enemy should cover—

- Size and/or strength.
- Actions or activity.
- Location and direction of movement.
- Unit identification. (The designation of the enemy unit may be derived from unit markings, uniforms worn or through prisoner interrogation.)
- Time of observation.
- Equipment and weapons.

Messages are printed in block letters. Individual items of information are numbered and separated into paragraphs. If doubt exists as to message receipt by the commander, a summary of its contents is included in the next message. Information about the originator or writer should cover—

- Location at the time of enemy observation (reference to an important terrain feature, by map coordinates, by the back azimuth from each of two definitely located points, or the back azimuth and distance from one known point).
- Intentions. (Remain in position? Continue on the mission? Take other action?)

The message is carefully reread and if possible read by another person to ensure understandability. If a messenger is used, the messenger must read and understand the message in order to answer any questions the commander might have.

Sketch

Information difficult to describe may be given accurately on a simple sketch. The sketch may give all the necessary information or it may be used to supplement a written message. A military sketch is generally one of two types: simple or panoramic. The simple sketch is easily made and read.

Figure 6-4 shows a simple sketch that has been included in the message itself. The sketch may be on a separate sheet of paper, but all of the necessary information must be contained in either the sketch, the message or both.

The panoramic sketch is a picture of the terrain's elevation in perspective, as seen from one point of observation. Although a panoramic sketch is not difficult to create, skill and training are necessary to enhance usefulness. Figure 6-5 on page 6-4 illustrates how to make a panoramic sketch.

Overlay

The same information sent back to higher headquarters on the sketch may be sent on an overlay, if the sender and the person to whom the message is to be sent have copies of the same map. Figure 6-6 on page 6-6 illustrates a simple overlay. The overlay is drawn on transparent paper as follows:

- Orient the map and place it on a hard, flat surface.
- Place the transparent paper over the part of the map of the object or information to be transmitted and hold the paper in this position.
- Orient the overlay to the map by tracing in the intersecting grid lines at two opposite corners of the overlay. Write the correct number designation on the overlay. The cross made by the intersection is called a *tick mark* and enables the receiver to locate the exact area on the map covered by the overlay.
- Sketch the object seen or the information to be transmitted on the tracing paper (the sheet on top of the map) in the exact location it would appear on the map (the sheet underneath the tracing paper).

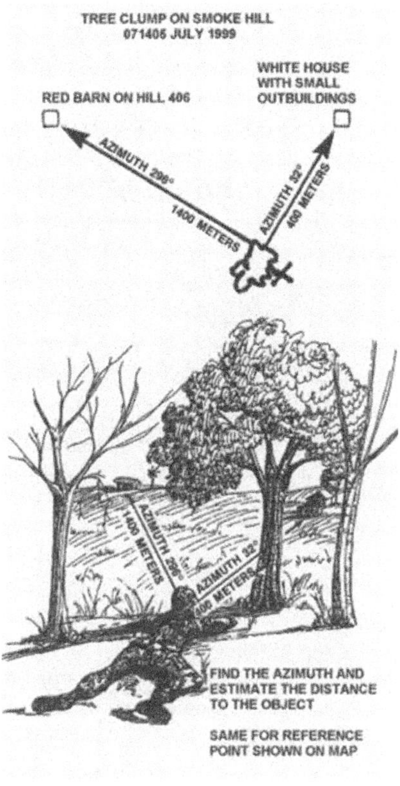

Figure 6-4. Making a Simple Sketch.

Explanatory notes are annotated in the overlay's margin, arrows point to the objectives.

- Indicate with an "X" and an appropriate explanation the position from which the observer saw the object or obtained information.
- Include title and scale of the map from which the overlay was made, date and hour the information was obtained, and signature of the observer on the overlay in the lower right-hand corner.

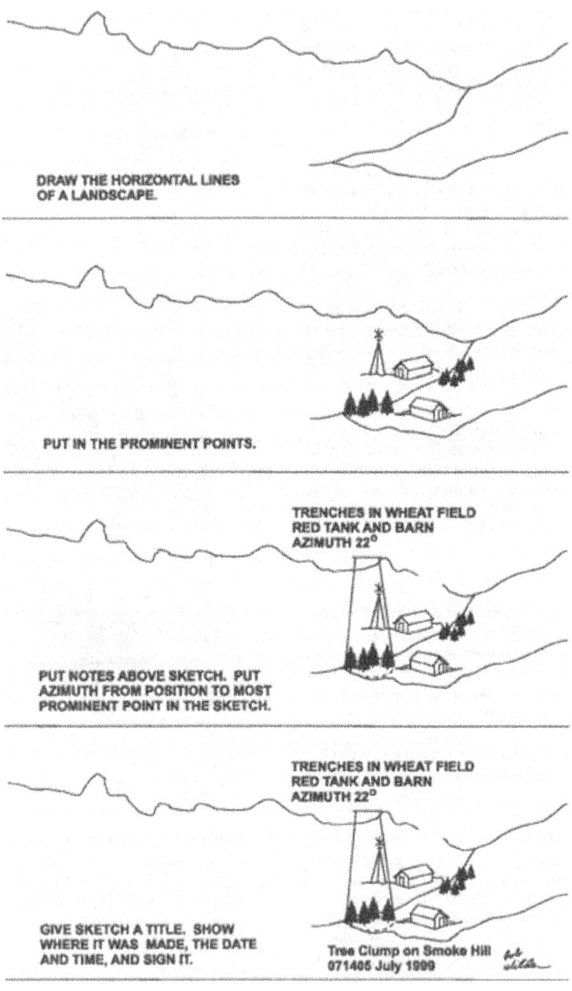

Figure 6-5. Panoramic Sketch.

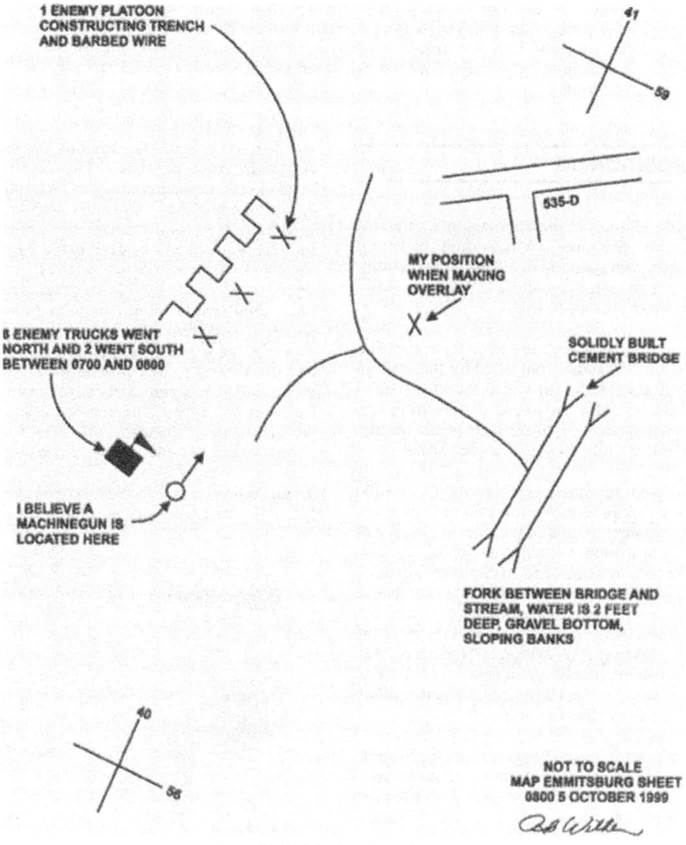

Figure 6-6. Simple Overlay.

CHAPTER 7. SCOUTING FIRE TEAMS

Upon completion of individual training, a Marine adapts this training to the functioning of the unit. The basic tactical unit for scouting is the fire team. Within the fire team, scouts normally work in pairs to furnish security and gather information. Information gathered by the individual scouts is passed to the fire team leader.

7001. POSITIONING

When a rifle platoon in the approach march is not preceded by adjacent units (left, right, front, and rear), it employs its own scouting elements. The scouting element for a platoon is usually one fire team; however, an entire squad may be used.

A fire team used as a scouting element is called a scouting fire team and is controlled by the platoon commander, assisted by the squad leader. A squad leader whose squad is providing the scouting fire teams normally marches near the platoon commander to assist in the control of the scouting fire teams.

A scouting fire team moves aggressively to cover the front of the advancing platoon and to locate the enemy's position(s). It generally moves in a wedge or skirmisher's formation. Normally, a scouting fire team is deployed on a frontage of 50 to 75 meters (10 to 17 meters between each individual scout). The entire squad may be employed to cover a wider frontage. The platoon commander coordinates the movement of the scouting fire team(s) so as to protect the main body of the platoon from enemy fire from points within 400 to 600 meters away, or in close terrain from points within the limits of enemy observation.

Scouting fire team(s) should have enough firepower to overcome resistance from small enemy advanced posts and patrols; the intent is to make enemy riflemen and machine gunners open fire and disclose their positions. Without scouts in advance, the platoon may move into areas where enemy fire may prevent further advance or maneuver and inflict heavy casualties (see fig. 7-1).

Scouting fire teams are covered by the platoon or, when the platoon is masked, the fire team leader ensures individual scouts are maneuvered and coordinated so that the fire team covers its own advance. The fire team leader constantly watches for signals from the platoon commander and remains in visual contact at all times.

The distance between the scouting fire team and the platoon is terrain-dependent. The scouting fire team should not be beyond visibility of the platoon. In open terrain, the platoon commander usually directs the scouting fire team to move by bounds along a succession of locations designated by the platoon commander as intermediate objectives.

Individual scouts should advance as stealthily as possible, while remaining consistent with their mission of reconnaissance to the front, taking advantage of cover without delaying the advance. An occasional glimpse of scouts constantly advancing over a wide front can make the enemy uneasy. It is this activity, and not the target the scouts offer, that may cause the enemy to open fire and disclose its location.

When fired upon, scouts must drop to cover and return fire only when necessary to complete the mission. If

Figure 7-1. Position of Scouts Preceding an Attacking Platoon.

necessary, one scout reports back to the fire team leader who informs the platoon commander.

7002. LOCATING ENEMY POSITIONS

The key terrain of defensive positions are those points that afford extended observation over the ground where the attack must advance. The enemy will place machine guns and infantry to defend critical points.

Members of a scouting fire team preceding an attacking platoon identify the probable enemy infantry and machine gun positions (see fig. 7-2). They use concealment and cover to conduct their advance in order to discover the exact location of enemy positions.

7003. ACTION WITH AN ATTACKING PLATOON

The scouting fire team reconnoiters to the front of the advancing platoon. As soon as the scouting fire team leader indicates the area is secure, the platoon advances and the scouting fire team moves forward. Squads within the platoon advance by bounds; at least one squad is positioned to support the other(s) by fire. Successive positions along the line of advance are selected and designated by the platoon commander as intermediate objectives, and reconnoitered by the scouting fire team before occupation. By conducting proper reconnaissance, surprise by the enemy or movement in the wrong direction may be prevented.

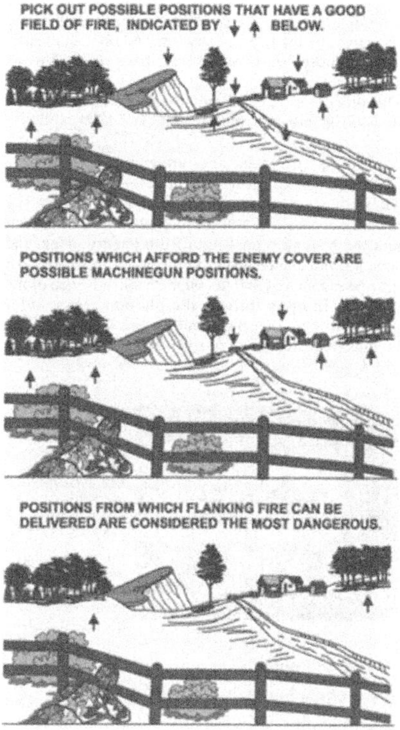

Figure 7-2. Assessing Probable Enemy Machine Gun Positions.

Movement

The distance between the scouting fire team and the front of the main body of the platoon is dependent on the mission, enemy, terrain and weather, troops and support available, and time available (METT-T). In close terrain, such as dense woods, the scouting fire team's movements closely resemble those used for night operations. In approaching houses, woods, and villages, one scout of each pair covers the other while the latter reconnoiters (see fig. 7-3).

A scouting fire team moves forward aggressively to cover the front of the advancing platoon, usually adopting either skirmisher's or wedge formation in order to be prepared to go into action immediately and to cover a wide frontage of 50 to 75 meters.

As a scouting fire team advances in open terrain, it is supported when possible by elements of the platoon; in close terrain, by mutual support within each fire team. Mutual support within the fire team is accomplished by the fire team leader and the automatic rifleman forming a team that supports by fire the advance of the rifleman and assistant automatic rifleman until they reach a location designated by the fire team leader. The rifleman and assistant automatic rifleman then support by fire the movement of the team leader and automatic rifleman. These successive points to which the sub elements of the fire team move are normally designated as fire team intermediate objectives by the fire team leader. The fire team leader sets as many fire team intermediate objectives as necessary to maintain mutual support within the team. This process is repeated until the team can be covered by other elements of the rifle platoon.

A scouting fire team takes advantage of available cover and concealment without delaying its advance. The orders of the platoon commander govern the distance at which it precedes the platoon. The terrain and the probable position of the enemy affect the scouting fire team's distance in front of the platoon. It may be as much as 400 to 600 meters in advance of the platoon. In open terrain, the platoon commander usually directs that the scouting fire team move by bounds to a succession of intermediate objectives. In

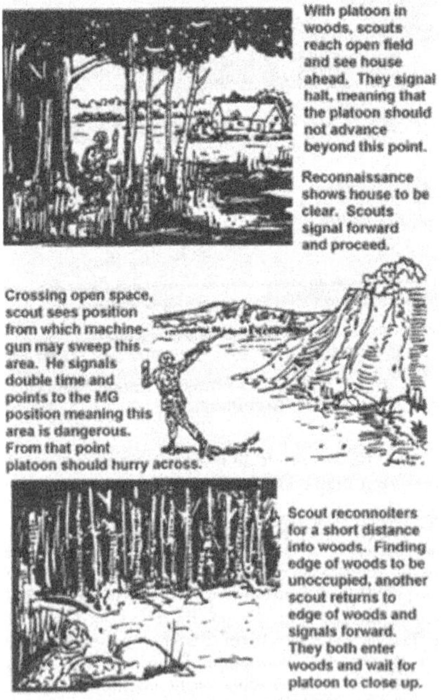

Figure 7-3. Conduct of Scouts Preceding a Platoon.

close terrain or conditions of limited visibility, the scouting fire team is normally ordered to precede the platoon at the limit of visibility, maintaining visual contact with the platoon commander.

Action in Woods

When a scouting fire team is directed to advance over open ground to the edge of a woodline, two members of the team, preferably the rifleman and assistant automatic rifleman, reconnoiter inside the woodline while the remainder of the fire team covers them. It is not recommended they separate until finished with their reconnaissance of the far side of the danger area. Both members staying together can cover the same area using a zigzag reconnaissance and they are better equipped to overcome any opposition.

In heavy underbrush and/or poor visibility, the rifleman and assistant automatic rifleman proceed into the woodline together for 50 to 60 meters. The two then separate, searching out either flank to the first high ground or limits of observation, probably 50 to 100 meters (see fig. 7-4 on page 7-4). After the initial search and out posting the limit of advance, the remaining scout signals the fire team forward.

In light underbrush and/or good visibility, the assistant automatic rifleman remains at the edge of the woodline while the rifleman searches the woodline. (see fig. 7-5 on page 7-5). The rifleman searches the woodline in a zigzag pattern, reports back to the assistant automatic rifleman, then moves to an outpost position at the limit of advance. In turn, the fire team leader signals the platoon commander that it is safe for the platoon to move forward.

The fire team leader then moves the remainder of the fire team into the woods, joining up with the forward scout manning the outpost. The scouting fire team occupies and holds a line 50 to 75 meters within the woods and observes toward the direction of movement until the platoon closes up. The scouting fire team leader awaits further word from the platoon commander before moving the team further into the woods.

When directed, the scouting fire team leader moves the team forward until they reach the far edge of the woods. The team is held at the edge of the woods and the fire team leader notifies the platoon commander of the situation. The platoon commander moves the platoon to a position where it can cover the scouting

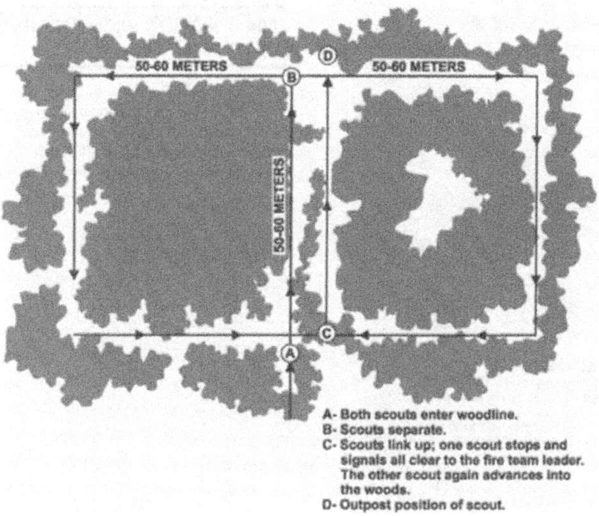

Figure 7-4. Searching Edge of a Woodline (Dense Underbrush/Poor Visibility).

fire team as it exits the woods and directs the team leader to move out and continue the scouting mission.

A scouting fire team passing through woods ahead of its unit maintains a distance allowing visual and oral communications. If an obstacle is encountered, reconnaissance to its front and flanks must be carried out. When advancing along a road or path, scouts precede the platoon to provide necessary protection and to prevent surprise fire on the platoon. When crossing a road or path, they reconnoiter well to the flanks before signaling "all clear" to the platoon.

The scouting fire team will not exit the woods until the arrival of the platoon commander, who will then be given an opportunity to alter the disposition or direction of march. The point where the platoon exits the woods is considered a danger area where the platoon is vulnerable to enemy fire. The scouting fire team is sent ahead to reconnoiter the danger area, as well as the next area to be occupied by the platoon. They signal back whether conditions require a halt, an advance or a quick rush across the open area. The scouting fire team leader must be continually on the lookout for signals from the rear.

Action Under Fire

When a scouting fire team is fired upon, they immediately take cover, locate targets, and return fire. The scouting fire team leader then determines—

- Location of enemy (range and reference points).
- Extent of position (location of flanks).
- Types of positions (obstacles, bunkders, fighting holes, etc.).
- Number of enemy.
- Enemy weapons (machine guns, mortars, tanks, etc.).

The platoon commander assesses the situation as quickly as possible based on the limited information obtained. Usually the platoon commander brings up the remaining squads, sets up a base of fire, and assaults the enemy position. Should the enemy position prove too strong for the platoon, the platoon remains engaged with the enemy as a base of fire until the remainder of the company is committed to clear the enemy resistance.

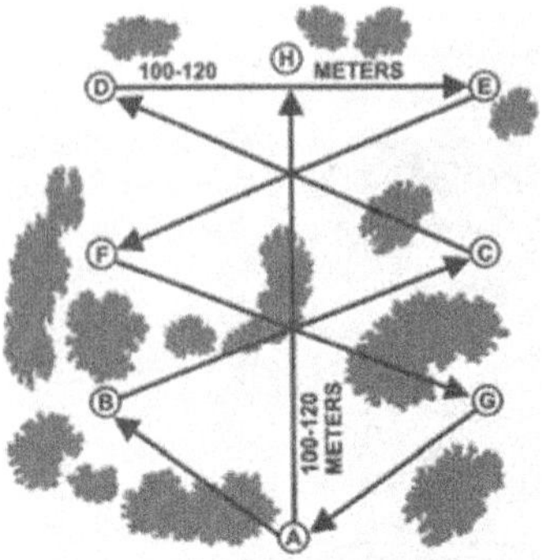

A- Both scouts enter woods. Assistant automatic rifleman remains here and covers forward movement of rifleman.

B through G - Rifleman moves to these positions, stopping and observing before moving to next position.

G to A - Rifleman reports results of his search to the assistant automatic rifleman.

A to H - Rifleman moves to out-post position.

Figure 7-5. Searching Edge of a Woodline (Light Underbrush/Good Visibility).

7004. ACTION WITH AN ENVELOPING UNIT

When a platoon is given the mission to envelop an enemy position, a scouting fire team is employed for protection and reconnaissance in the same way as when the platoon is advancing in the approach march.

PART 2. INFANTRY PATROLLING

CHAPTER 8. FUNDAMENTALS OF INFANTRY PATROLLING

This chapter begins Part 2 and provides basic information about infantry patrols; specifically, their purpose, types, and missions. Infantry patrol training is also addressed and keys to successful patrolling are presented. Subsequent chapters of this part cover patrol organization, preparation, movement, and reconnaissance actions.

8001. DEFINITIONS

A patrol is a detachment of ground, sea or air forces sent out for the purpose of gathering information or carrying out a destructive, harassing, mopping-up or security mission (Joint Publication [JP] 1-02). The mission to conduct a patrol may be given to a fire team, squad, platoon or company.

8002. RELATION OF PATROLLING TO SCOUTING

Each patrol member must be knowledgeable in the principles of scouting and maintain membership of a larger team. To develop the teamwork skills required among the members of a patrol, additional training beyond the basic principles is necessary to become a well-trained scout. A patrol member must respond quickly to the decisions and orders of the patrol leader. There must be complete confidence among all members of the patrol and the confidence that they, as a team, will be successful in their mission.

8003. PURPOSE

A commander must have current information about the enemy and the terrain in order to employ the unit effectively. Patrols are an important means of gaining this information and are used to destroy enemy installations, capture enemy personnel, perform security missions or prevent the enemy from gaining information. Modern warfare places a high premium on effective patrolling because units have larger areas of operations and can be threatened from all directions. As distances between units increase, more patrolling becomes necessary to prevent infiltration by guerrillas or small enemy units, as well as to maintain contact with friendly adjacent units. Active patrolling by numerous small groups is needed to locate the enemy and gather information on the enemy's disposition, strength, morale, and weapons, as well as gather and confirm information about the terrain.

8004. TYPES OF PATROLS

Classification as to Mission

Reconnaissance Patrol

Reconnaissance patrols gather information about the enemy, terrain or resources. Relying on stealth rather than combat strength, they gather this information and fight only when necessary to complete the mission or to defend themselves. The distance covered by reconnaissance patrols varies based on the terrain and mission. The squad is ideally suited for reconnaissance patrol missions because of its relative small size and its experience of working together.

Combat Patrol

A combat patrol is a fighting patrol assigned missions that require engagement with the enemy in combat. Larger and more heavily armed than reconnaissance patrols, combat patrols have a mission to capture enemy documents, provide security, and capture or destroy enemy equipment and installations. Such action is ordinarily followed by a return to friendly positions. Regardless of the mission, the patrol reports any information concerning the enemy and terrain acquired during the accomplishment of the assigned mission. There are four types of combat patrols: raid, contact, ambush, and security (normally conducted by a Marine rifle platoon). A rifle platoon reinforced with crew-served weapons is normally considered the minimum size for contact, economy of force or ambush patrols. In some situations, such as the capture of a small enemy outpost, a rifle platoon could conduct a raid. However, a raid is a complex mission and, due to the organization of a raid force (command, reconnaissance, assault, support, security, and reserve elements), a rifle company is normally the smallest force assigned to a raid.

Classification as to Means of Movement

Foot Patrols

Movement by foot is the most common means; however, there are inherent disadvantages. Foot patrols travel slowly and carry limited quantity and types of equipment and supplies. Range and area coverage is relatively restricted. Foot patrols also have apparent advantages in that they have fewer restrictions as to terrain that can be covered; are more difficult for the enemy to detect; provide thorough coverage within limits of range; and are generally not inhibited by weather.

Motorized Patrols

Where terrain and road networks permit, a motorized patrol overcomes the inherent disadvantages of the foot patrol. Mechanized forces require patrolling units that can keep pace with them. However, motorized patrols are restricted to certain types of terrain, and tend to bypass areas that may be advantageous to and occupied by enemy infantry.

Waterborne Patrols

Waterborne patrols move over seas, lakes, rivers and streams, canals, and other inland waterways. The water is either used as a medium of entry to an objective area or is the actual patrol route. Waterborne patrols are limited by the location of water routes in the terrain and tend to bypass areas that may be advantageous to and occupied by the enemy.

Helicopterborne Patrols

Where terrain is extremely difficult or the enemy situation precludes the use of vehicle or motorized patrols, helicopterborne patrols are a method or means to conduct a patrol.

8005. TRAINING

Training is essential to successful patrolling. Premature and unordered actions by members of the patrol destroy coordination and control. Leaders are trained to issue their orders calmly to inspire confidence and discipline, and to avoid misunderstanding. Patrol members must work together and fight as a team. Training should develop the following skills:

- Expertise in handling individual and special weapons, and familiarity with enemy weapons that may be captured.
- Recognize camouflaged personnel, equipment, and defensive positions; ability to pick up fleeting targets fire the rifle from any firing position.
- Understand fire discipline and, after weapon firing, immediately change position (see fig. 8-1).
- Quick and accurate observation skills, and the ability to recall and transmit clearly and briefly, both orally and in writing.

Figure 8-1. Changing Position After Firing.

- Recognize and quickly respond to improvised signals, visually or by sound.
- Ability to swim with weapon and equipment.
- Use issued or improvised camouflage suits and garnish helmet in order to blend with the surroundings. Smudge face, hands, and any bright surfaces of weapons and equipment with some substance, such as mud or charcoal, to prevent the reflection of light.
- Silence self, equipment, and weapon.
- Use antimalarial and water purification tablets.
- Acclimation to temperature extremes.
- Develop a sense of direction and learn how to follow a course by compass, stars, sun, flow of streams, prominent terrain features, and by observing other natural phenomenon. Learn to determine the distance traveled from a known point and to keep a record of azimuths and the distance traveled on each azimuth (dead reckoning).
- Call for and adjust indirect fire assets.
- Familiarization with all communications assets and the use of field expedient antennae.

8006. KEYS TO SUCCESSFUL PATROLLING

Regardless of the category or means of conducting a patrol, the keys to successful patrolling are—

- Detailed planning. Every portion of the patrol must be planned, all possible contingencies considered.
- Productive, realistic rehearsals. Each phase of the patrol is rehearsed, beginning with actions in the objective area. Similar terrain and environmental conditions are used when conducting rehearsals.
- Thorough reconnaissance. Ideally, the patrol leader will physically conduct a reconnaissance of the route and objective. Photographs and/or maps will be used to supplement the reconnaissance.
- Positive control. The patrol leader must maintain positive control, this includes supervision during patrol preparations.
- All-around security. Security must be maintained at all times, particularly near the end of the patrol where there is a natural tendency to relax.

CHAPTER 9. PATROL ORGANIZATION

Organizing a patrol is a two-step process: the general organization of the entire patrol and the task organization of various patrol elements. Normally, the nature of patrolling does not permit long preparation periods and rehearsals to specifically build a unit for each mission. Accordingly, the patrol leader must combine unit integrity considerations with proven concepts of patrol organization.

9001. GENERAL ORGANIZATION

The patrol leader establishes a patrol headquarters and elements to accomplish the mission.

The headquarters is composed of the patrol leader and the personnel who provide support for the entire patrol, such as a forward observer, corpsman, and radio operator.

The major subdivisions of reconnaissance and combat patrols are elements. The existing infantry structure (squads and fire teams) is reinforced as required. For example, a reinforced platoon tasked to conduct a combat patrol that will raid an enemy outpost could be organized as follows:

- Platoon headquarters (command element):
  - Patrol leader (platoon leader).
  - Assistant patrol leader (platoon sergeant).
  - Navigator.
  - Radio operator (company tactical net).
  - Radio operator (patrol tactical net).
  - Corpsman.
- The first squad (security element) provides security en route to the objective area (the point, flank security, and rear security) and at the objective area (the flanks and the objective rally point).
- The second squad (support element) supports by fire for the attack, covering fire for the withdrawal, and supporting fires to cover the crossing of danger areas.
- The third squad (assault element) provides the assault force to attack and seize the objective; searchers to clear the objective; pacers, compass man, navigator, and the assistant patrol leader en route and back from the objective area.

Any attachments a patrol may have (i.e., demolition team, scout snipers, and machine gun squad) will be added to the element that supports its function. For example, the demolition personnel should go with the unit conducting the attack, and scout snipers and machine gunners should stay with the support squad.

9002. TASK ORGANIZATION

The preceding paragraphs described the elements necessary for a patrol to accomplish its mission. These elements reflect the internal functions or tasks required for the patrol to succeed. Depending on the METT-T, there are various methods of grouping these elements together. Task-organization is the further subdivision of patrol elements into teams that are required to perform essential tasks. In creating teams, unit integrity of infantry units should be maintained.

The patrol is organized so each individual, team, and element is assigned a specific task, but capable and prepared to perform other tasks. This may not be possible for certain specialist tasks requiring a trained technician.

CHAPTER 10. PATROL PREPARATION

For a patrol to succeed, all members must be well trained, briefed, and rehearsed. The patrol leader must have a complete understanding of the mission and a thorough understanding of the enemy and friendly situations. The patrol leader should make a complete reconnaissance of the terrain to be covered (either visual or map), and must issue an order to the patrol, supervise preparations, and conduct rehearsals.

10001. MISSION

The mission assigned to a patrol must be clear and oriented toward one objective with a specific task and purpose. More than one primary objective or indefinite missions invites confusion, casualties, and failure.

10002. FACTORS INFLUENCING PATROL SIZE

The circumstances under which patrolling by infantry units is conducted make it necessary for combat patrols to be able to fight offensively, security patrols to defend themselves, and reconnaissance patrols to move quickly and only fight if necessary to break contact or defend themselves. The size of a patrol depends on METT-T.

Generally, a patrol should contain the least number of members needed to accomplish the mission. Combat missions ordinarily require larger patrols than reconnaissance missions.

10003. COMMANDER'S DUTIES

Determining Patrol Requirements

The need for conducting patrols derives from the commander's stated mission (issued by higher level commander) and other specified and implied tasks (secondary or supporting efforts necessary to accomplish the stated mission). This requires analyzing the unit's mission and determining the necessary reconnaissance and/or combat tasks that must be performed. By considering the mission, along with time available, the commander develops the overall concept of operations to include the patrol plan and the specific tasks for each patrol.

Assigning Units

When assigning patrol missions, maintaining the integrity of the existing unit while considering the skills and experience of the unit and its leader are critical factors to the infantry commander. To provide operational depth and equitable apportionment of hazardous assignments, the commander ensures that each of the subordinate leaders and units develops the skills and experience necessary to conduct successful patrols. Assignment of patrol units must consider the commander's concept of operations as a whole and the plans for subsequent employment of assigned forces after completion of the assigned patrolling mission.

Providing Adequate Time

The commander must allow the patrol sufficient preparation time by completing the mission analysis, estimate, and preparation of orders. The commander should use warning orders to alert subordinates to possible requirements and afford them the opportunity for concurrent planning.

Providing the Patrol Leader Information

The following information should be provided:

- A simple, straightforward explanation of the mission, particularly for night patrols.
- General routes (defined by checkpoints) or exact routes (defined by avenues of approach or other terrain features) to follow.
- Enemy composition, disposition, and strength.
- Location and activities of friendly troops.
- Outposts or other security elements through which the patrol is to pass.
- Terrain conditions.
- Missions and routes of other patrols.

- Time patrol is to depart and return.
- Method of reporting information while on patrol (radio, messenger), place where messengers are to be sent, and place where the patrol leader is to report upon completion of the patrol.
- The challenge and password to be used for exiting and reentering friendly lines.
- Special instructions such as locations to be avoided and essential elements of information higher headquarters is seeking.
- Report barrier/obstacle plan of friendly units if known; support available from friendly unit (e.g., medical evacuation [MEDEVAC], react force).
- Fire support available.

Providing Special Skills and Equipment

There will be situations when the unit assigned to conduct a patrol does not have the necessary technical skills or equipment organic to it to successfully accomplish the patrol mission. In these cases, additions (attachments) are made to the unit, such as—

- Machine gun and/or shoulder-launched, multipurpose assault weapon (SMAW) team and/or squad.
- Forward observers (mortar/artillery).
- Radio operator(s).
- Combat engineers.
- Tracked vehicle crewman to assess the ability to traverse the terrain.
- Corpsman.
- Other personnel (snipers, translators) as required.

Providing Miscellaneous Support

The commander must ensure that the patrol leader is provided with the food, water, ammunition, radios and batteries, maps, special clothing, and any other items required by the unit (includin attachments) needed for the mission. Post-patrol support such as debriefings by intelligence personnel must also be planned.

Reviewing the Patrol Leader's Plan and Preparations

Once the patrol leader receives the mission, conducts visual and/or map reconnaissance, and develops the plan, the commander confirms the patrol leader's understanding of the mission and plan for accomplishing it. This discussion between the patrol leader and commander ensures that the patrol leader understands the commander's intent and is properly prepared to provide the patrol leader an opportunity to ask for clarification or additional support, if required.

Debriefing the Patrol

Upon return of the patrol, the commander receives the patrol report at a debriefing attended by the patrol leader and all patrol members. The debriefing should be conducted as soon as possible following the patrol's return, while information is still fresh in the minds of the patrol members. A patrol report, based on the information collected during the debrief, is generated and forwarded to the next higher com-mander.

10004. PATROL LEADER DUTIES

The patrol leader organizes and prepares the patrol by using the six troop-leading steps to make the best use of resources available. These steps are to—

1. Begin planning.
2. Arrange for reconnaissance and coordination.
3. Make reconnaissance.
4. Complete the plan.
5. Issue the order.
6. Supervise.

Begin Planning

The patrol leader begins by evaluating all factors affecting the mission. He looks for possible courses of action that lead to a decision, and then transforms this decision into an order. The first step includes making an initial assessment and decision on using available time, issuing a warning order and initial preparatory tasks, and initiating his estimate.

The initial planning effort assesses the time, assistance, and information available, and plans the proper use of each. Time allowances include reconnaissance; completion of the estimate and order; troop preparation; and such briefings, rehearsals, and inspections as required before beginning the patrol.

The patrol leader reviews the mission and the attachments and/or support available and decides what preparatory efforts must begin immediately. Proper use of subordinates to manage these initial tasks during this period reduces preparation time and frees the patrol leader for proper planning and reconnaissance. A warning order is issued using a modified five-paragraph order format (situation, mission, execution, administration and logistics, and command and signal [SMEAC]) as a checklist. (Refer to app. A for specific information that a patrol warning order should contain within each checklist item.

Once the initial preparations are set in motion, the patrol leader begins estimating by analyzing the mission; considering the friendly, enemy, and operating environments; considering each course of action available against what the enemy might do; comparing the courses of action in terms of mission accomplishment, capabilities, and probable casualties; and choosing one that becomes the basis for concept and order. While the patrol leader should use and organize notes, estimates must be done quickly and accurately, particularly for immediate situations. To organize thinking, the patrol leader uses METT-T, which consists of:

- Mission—the mission assigned to the patrol and how it relates to the mission of the commander who is sending the patrol.
- Enemy—what is known or suspected of enemy presence and capabilities, habits and characteristics, and fighting techniques.
- Terrain and weather—including ground, vegetation, drainage, weather, and visibility.
- Troops and support available—friendly situation and support available.
- Time available—the constraints and impact of time on preparation and mission accomplishment.

The estimate begins with mission analysis, which is the most important part of the entire planning process. Here, the patrol leader considers the specified tasks of the mission assigned and identifies other significant actions (specified and implied tasks) that must be undertaken to accomplish the stated mission. These tasks are arranged in sequence of accomplishment. The tasks and sequence create the framework for developing courses of action for the patrol concept of operations.

Arrange for Reconnaissance and Coordination

The patrol leader arranges a personal reconnaissance to observe as far forward as possible and also coordinates with the appropriate commanders for the patrol's "passage of lines" (see paragraph 11001 for complete definition) and supporting fires. The patrol leader also coordinates with other patrol leaders who may be operating in the same or adjacent areas and requests that the commander assigning him the patrol mission coordinate the patrol action with adjacent commanders, local security, and night defensive fires, as appropriate. The patrol leader may delegate any or all of these arrangements to the assistant patrol leader if the patrol leader requires the time for planning.

Make Reconnaissance and Complete the Estimate

The patrol leader uses personal reconnaissance to answer questions that arise from the map reconnaissance and METT-T evaluation. Specific points include passage points, lanes through obstacles, locations of friendly listening posts and observation posts, possible approach and return routes, enemy positions (if any), and intermediate observation points on the way to the objective.

In selecting approach and return routes, the patrol leader chooses routes that best use concealment and avoid opposition and obstacles. To lessen the chances of ambush by the enemy, the return trip is planned along a different route. In addition to personal reconnaissance and review of the map and aerial photographs, the advice of other patrol leaders who already are familiar with the terrain and the objective area should be considered.

After compiling information about the situation and possible time constraint, the patrol leader completes an estimate. The first step is developing courses of action, each of which will provide for movement to the objective area, mission accomplishment, and the return, based on the tasks and their sequencing identified in mission analysis. While the eventual concept of operations is presented in order of occurrence, the patrol leader must develop the courses of action by either backward or forward planning. In situations where the objective is well defined and there is sufficient information to plan the action for mission accomplishment (reconnaissance or combat), the patrol leader begins the scheme for accomplishing the mission at the objective and then, planning backwards, considers the options for getting there and back.

The following sections on movement to and return from the objective area, reconnaissance missions, and combat missions discuss methods and options available to the patrol leader in developing the courses of action. The principal variables between courses of action will be who, where, and how in the following:

- Patrol task organization.
- Routes to the objective area.
- Observation point(s) (reconnaissance patrol), ambush site, form of maneuver, type of ambush (combat patrol), fire support plan.
- Return routes.

The patrol leader then mentally considers the progress of each course of action (a map or simple sketch is a useful aid) against expected and unexpected enemy action. By comparing the options against each other and prospective enemy opposition, the patrol leader chooses the course of action that has the best chance of success. Includes in this mental preview process is the time to determine the patrol plan for unexpected contingencies (enemy attack/counterattack, casualty handling). These contingency actions, together with the selected course of action, become the patrol concept of operations.

Once the patrol leader determines the scheme of maneuver, the fire support required to accomplish the mission is addressed including the fire power organic to the patrol and what additional indirect fire support will have to be provided by other units. When planning for indirect fire support, the patrol leader considers the following questions:

- Will artillery, mortar, or close air support be required at the objective area (combat patrols)?
- What artillery and mortar targets exist along the routes to and from the objective area that can be employed by the patrol if it encounters the enemy during movement (reconnaissance patrols and combat patrols)?
- What additional fire support will be required to cover the patrol's movement from the objective area back to the friendly area once the enemy is aware of the patrol's actions at the objective area (combat patrols)?

The effect that casualties have upon the patrol depends upon many factors. Generally, more casualties can be expected in a combat patrol than in a reconnaissance patrol. A patrol may continue on to the objective carrying its casualties, send them back with a detail of Marines, abort the mission and return the entire patrol with the casualties, or call their parent unit for assistance.

Some factors that determine what action the patrol leader takes are: patrol's mission; unit's standing operating procedure for handling wounded; number of casualties and nature of their injuries; availability of aid, helicopters or other means of casualty evacuation. Helicopter evacuation should only be used for the most serious casualties. For infantry units conducting patrols in proximity to the enemy, helicopter evacuation of casualties may compromise the patrol's mission and force the patrol to return to friendly positions before the mission is completed.

The patrol leader determines the requirement for nuclear, biological, and chemical (NBC) defense equipment. Gas masks should always be carried due to the availability of riot control agents (RCAs) to the enemy. If chemical or biological agents have been employed in the area that the patrol must pass through, protective garments will have to be worn by patrol members for part of or the entire patrol. Wearing extra clothing and carrying extra equipment affects the speed of the patrol's movement. A contingency plan for post patrol decontamination must be developed.

Complete the Plan

At this point, the patrol leader has completed the basic thinking necessary for accomplishing the assigned mission. The patrol leader prepares the patrol order to spell out the details, assign tasks to subordinates, and explain the entire endeavor for ease of understanding by the other members of the patrol.

Prepare the Order

The patrol leader's order contains more detailed information than discussed in the warning order. Orders follow the prescribed five-paragraph order format but contain greater detail. The patrol order is a modified 5-paragraph order; the major modifications are to paragraphs 3a and 3c. The format for the order is contained in appendix B.

Issue the Order

The patrol leader asks for a status report on the initial preparatory tasks assigned to subordinate leaders and specialists when the warning order was issued. When the patrol leader has completed planning and initial preparations have progressed to the point where the patrol order may be issued, the members of the patrol are assembled. Roll call is taken to ensure all patrol members are present, then the prepared order is issued.

This will be the only opportunity for the patrol leader to issue detailed instructions. The mission, in particular, must be unmistakably clear so that once the patrol is committed, all subordinate leaders can act with unity of purpose.

Whenever possible, the patrol leader should have a Marine, such as the navigator, build a terrain model using dirt, sand, twigs, etc., explaining the concept of operations for movement to the objective area, actions at the objective area, and the return. Terrain models provide patrol members with a clear and simple layout of the area of operations and key terrain.

Supervise

Inspections and rehearsals, vital to proper preparation, are conducted even though the patrol leader and patrol members are well experienced in patrolling. Inspections determine the patrol's state of physical and mental readiness.

The patrol leader inspects before rehearsals to ensure completeness and correctness of uniform and equipment. The following areas are checked:

- Camouflage.
- Identification tags, Geneva Convention cards.
- Prescribed equipment, weapons, and ammunition are available and serviceable.
- Tape and other items are used to "silence" equipment (prevent noise produced during movement).
- Items that could provide information to the enemy (e.g., letters and papers) remain behind.
- Unnecessary equipment and excess weight remain behind.

The patrol leader questions each patrol member to ensure the following is known:

- The mission, planned routes (primary and alternate), and the fire support plan of the patrol.
- The individual's role: what to do and when to do it.
- What others are to do and how their actions impact.
- Challenges and passwords, codes, reporting times, radio call signs, frequencies, and any other pertinent details.

There is usually a period of time between final rehearsal and departure. The patrol leader reinspects just before departure to ensure all equipment is still in working order and the unit is ready to embark on the mission.

Rehearsals ensure the operational proficiency of the patrol. Plans are checked and needed changes are made. The patrol leader verifies the suitability of equipment. It is through rehearsals that patrol members become thoroughly familiar with the actions to take during the patrol.

If the patrol will operate at night, both day and night rehearsals are conducted. Terrain similar to that over which the patrol will operate is used. All actions are rehearsed. If time is limited, the most critical phases are rehearsed. Action at the objective is the most critical phase of the patrol and is always rehearsed.

An effective method is to talk the patrol through each phase, describing the actions and having each member perform individual duties. When satisfied, the patrol leader walks the patrol through all phases of the patrol using only the signals and commands to be used during the actual conduct of the patrol. Rehearsals continue until the patrol is thoroughly familiar with the plan. The rehearsal is also used to test the soundness of the patrol order and patrol organization.

After the rehearsal, the patrol leader makes final adjustments to the plan and patrol organization based on what was learned during the rehearsal and from other sources, such as the S-2 and adjacent patrols. When this is completed, the patrol leader issues final instructions to subordinate leaders noting any changes made in the patrol organization or plan. While the subordinate leaders are briefing the remainder of the patrol members, the patrol leader reports to the commander stating that the patrol is ready to begin the mission. The patrol leader also coordinates the location and time that the patrol can test fire all weapons prior to departure.

CHAPTER 11. MOVEMENT TO AND RETURN FROM THE OBJECTIVE AREA

This chapter provides guidance to patrol leaders for movement to and return from the objective area. Action in the objective area depends on whether the patrol is assigned a reconnaissance mission, combat mission or security mission. Chapters 12 and 13 provide detailed guidance on these types of missions.

11001. PASSAGE OF LINES

A passage of lines is an operation in which a force moves forward or rearward through another force's combat positions with the intention of moving into or out of contact with the enemy. (JP 1-02)

During the initial preparation for the patrol, the patrol leader selects a patrol assembly area and reconnoiters the area of passage designated by the commander. In coordination with the unit commander responsible for the area of passage, the patrol leader identifies gaps or lanes in minefields and wire obstacles and locates local security elements through which the patrol will pass. The patrol leader also checks the route from the patrol assembly area to the passage point or contact point where the patrol will depart friendly lines. If possible, both the route to the passage point and the route through the frontlines should be concealed from the enemy's view.

The patrol leader also reconnoiters the area for return passage of lines and coordinates with the unit commander responsible for the area of passage for passage points and lanes as necessary. The patrol leader observes these points from the direction that the patrol will use upon return to friendly lines, if possible, to aid in recognition upon return. The patrol leader provides the forward unit with information about the size of the patrol, general route, and expected time of return. The manner of challenge and recognition of the returning patrol should be coordinated in detail.

Upon return to friendly forward local security squads and/or frontlines, the patrol leader leaves the patrol in a covered position and moves forward with a radio operator and at least one Marine for security to make contact with the friendly unit in the manner agreed. After contact is made and recognized, the patrol leader rejoins the patrol and takes them to the passage point, personally checking in each member.

11002. ORGANIZATION FOR MOVEMENT

The patrol's task organization establishes the elements and teams needed to accomplish the mission in the objective area and to and from the objective. The patrol leader determines the formation(s) in which the patrol moves to the objective area.

Formations

The proper use of patrol formations is critical to the patrol's success. The squad and fire team formations described in FMFM 6-5 (proposed MCWP 3-11.2), *Marine Rifle Squad*, also apply to infantry patrolling formations. Because the movement of the patrol must be concealed from the enemy, the patrol normally moves through terrain that provides concealment. Control of the patrol in this type of terrain is difficult; thus the column formation, which is easily controlled, is normally used. However, as various types of terrain are encountered, the patrol leader uses the same considerations in determining the appropriate formation used in other infantry operations.

The standard squad and fire team formations are adaptable to a patrol. The patrol may change formations en route to match the situation and terrain. The patrol leader may have to sacrifice some control for better dispersion or give up some speed for greater stealth and more security. Other considerations include—

- Visibility, weather, terrain, and vegetation will influence dispersion and control of individuals and units. These factors may also affect the enemy; if visibility is good for the patrol, it is also good for the enemy. Two pieces of luminous tape worn on the back of the collar will aid in control and movement on dark nights. The collar is turned down when near the enemy. The tape can also be worn on the back of the cap, but should be covered or removed when near the enemy.

- Preserving the integrity of fire units (fire teams and, if attached, machine-gun teams) is of primary importance. If team members are detached from a machine-gun team, the unit integrity is lost and effectiveness is reduced. The patrol leader must also position fire units so as not to mask their fires.
- All-around defense of the patrol must not be sacrificed under any circumstances. The conventional squad and platoon formations provide adequate firepower in any direction required. When attachments are made to the patrol, the attachments are positioned within the formation to enhance the firepower of the patrol. If a fire unit, such as a machine-gun team or squad, is attached, it is incorporated into the all-around defense of the patrol by modifying the conventional formation (see fig. 11-1). It is permissible to employ machine-gun teams individually during movement. However, assault rocket squads and teams should be employed primarily as rifle fire units for movement. Firing the SMAW or AT-4 from within a patrol formation can present a serious back-blast danger to patrol members and should only be fired on the patrol leader's direction.
- Time allotted for mission accomplishment is also a major consideration. In selecting the formations, the patrol leader must consider the speed of the movement required to meet the time constraints (if any) imposed on the patrol. If required to meet a time schedule, a formation that permits rapid movement should be used. Speed, however, must never be permitted to force the patrol leader to make rash tactical decisions.

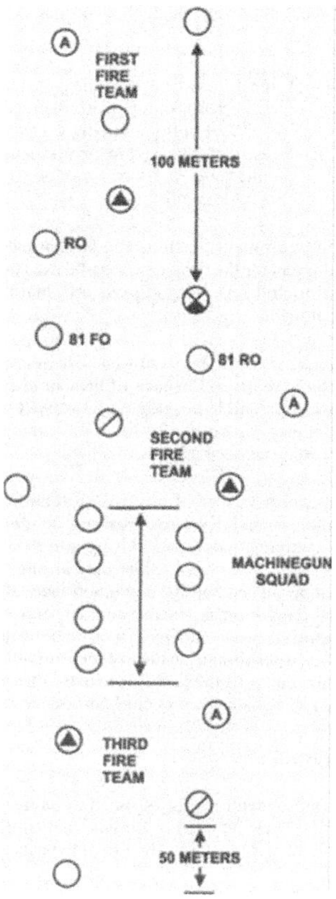

Figure 11-1. Squad Patrol with Attached Mortar Forward Observer Team and Machine Gun Squad.

Exercise of Control

The patrol leader is positioned for best control over the patrol. The assistant patrol leader moves at or near the rear of the patrol. Other subordinate leaders move with their elements. All patrol members assist by staying alert and passing on signals and orders. A signal to halt may be given by any patrol member, but the signal to resume movement is given only by the patrol leader.

Arm-and-hand signals are the primary means of communication within a patrol and should be used exclusively when near the enemy. All members must know the standard infantry signals (refer to FMFM 6-5), as well as any special signals required, and be alert to receive and pass them to other members.

The patrol leader should speak just loudly enough to be heard. At night, or when close to the enemy, the patrol leader halts the patrol and has subordinate leaders come forward. They speak in a low voice and then pass the information to their subordinates by moving from member to member.

Radios provide a means of positive control within a large patrol; however, radios should be used only when arm-and-hand signals or face-to-face contact between the patrol leader and subordinate leaders is impractical. When close to the enemy, words are spoken into the microphone with hands cupped over it in a low voice.

Other sound signals may be used if the patrol leader is sure they serve the purpose intended. Planned sound signals are rehearsed before starting on the patrol. Sound signals used must be simple, natural sounds that are few in number and easily understood. Bird and animal calls are seldom satisfactory.

Infrared equipment, such as the sniper scope and infrared filters for the flashlight, may be used as means of sending and receiving signals and maintaining control at night.

Luminous tape may be used to assist in control at night. Small strips on the back of the cap or collar of patrol members aid in keeping visual contact with the front member. However, the luminous tape must be covered when near the enemy.

An important aspect of control is accounting of personnel, especially after crossing danger areas, enemy contacts, halts, and exiting and re-entering friendly lines. The patrol leader may arrange for the last member to send up the count automatically after crossing danger areas, enemy contact, and halts. In large patrols or when moving in a formation other than a column, subordinate leaders check members and report the count to the patrol leader by the quickest method appropriate to the circumstances.

Navigation

One or more members are assigned as navigators for the patrol. Their function is to assist the patrol leader in maintaining direction by the use of the compass.

The patrol leader assigns at least two members as pacers to keep track of the distance from point to point. The average of their count is used to approximate the distance traveled. Pacers are separated so they do not influence each other's count. Pacers must know how to convert their own paces to meters.

The route is divided into increments, with each increment starting at a recognizable point on the ground. The pacers begin their counts from zero at the beginning of each leg. This makes the pace count easier to keep and checks for accuracy.

The pace count is sent forward when the patrol leader turns to the member behind and in a low voice says, "send up the pace" or uses the arm-and-hand signal of tapping his boot. This signal is passed to both pacers, who in turn send up the pace count in meters; for example, "two-hundred" or "one-seven-five."

Patrol members must understand that the counts of both pacers are sent forward. The patrol leader must know the counts of both men in order to check them.

Security

The patrol is dispersed consistent with control, visibility, cover, and concealment. Scouts are employed to the front, flanks, and rear of the patrol to provide security. For the patrol members traveling in the main body of the patrol, areas of responsibility are assigned to the front, flanks, and rear. Scouts are the eyes and ears of the patrol leader. They move when and where directed by the patrol leader and maintain contact with the patrol leader at all times, except when momentarily obscured by vegetation or other terrain features.

Front

Small patrols (squad size) may employ from one scout up to a fire team as the point, depending on the enemy situation, terrain, and route being followed by the patrol. Normally, squad-size patrols will use two scouts as the point. The point is responsible for investigating the route of advance immediately to the front of the patrol.

The point moves as far ahead of the patrol as visibility and terrain permit. When visibility is good, the point may precede the main body by as much as 100 meters. The point travels right and left ahead of the patrol, searching the area over which the patrol will pass.

The point maintains direction by knowledge of the general route to be followed and visual contact with the patrol leader. The patrol leader or the navigator ensures that the point is proceeding correctly.

The point, which stays far enough ahead of the patrol to provide security, is not a trail breaker for the patrol. If the point loses contact with the patrol, the point waits for the main body to catch up or moves rearward if contact is not quickly regained.

One of the navigators may be positioned with the point. One or more members works as the point while the other is the navigator.

Flank

Flank security for a patrol of squad size or less may be provided by using one or two members on either flank. If two scouts are assigned to a flank, one is positioned to observe the patrol leader and the other works farther out from the patrol. The scout who must observe the patrol leader remains within a maximum distance of 100 meters. The scout farther out remains in sight of the inside scout but normally does not move more than 20 or 25 meters away and remains prepared to relieve flank security regularly. Moving through dense woods or jungle may render the use of flank security impractical because of reduced visibility. In such cases, it moves with the patrol itself, but maintains observation to its assigned flanks.

Rear

A small patrol normally has only one rifleman assigned as rear security. An interval between the member assigned as rear security and the last Marine of the patrol is maintained at the limit of visibility, up to 50 meters. This member maintains rear security for the patrol by constantly observing to the rear.

Halting

Speed of movement is slower at night than in day patrols and reduces the danger of a Marine becoming separated from the patrol. The patrol occasionally halts to observe and listen for enemy activity; this is called a security halt. Upon signal, when reaching a danger area and periodically throughout movement en route, every member freezes in place, remains quiet, observes, and listens. It may be necessary to call a security halt just after departing friendly areas and just before entering friendly areas.

The patrol may halt briefly to send a message, eat, rest, check direction, or make a reconnaissance. The area selected should provide adequate concealment and cover, as well as favor the defense. All-round security is established and the patrol leader ensures all members move out when the patrol resumes movement. (For extended halts, see paragraph 11005.)

Infiltration

The disposition of enemy forces may sometimes prevent a patrol from entering the enemy occupied area as a unit; however, pairs of scouts or fire teams may slip through without being discovered. (Refer to FMFM 6-5 for infiltration techniques and procedures).

11003. CONTROL MEASURES FOR MOVEMENT

Checkpoints

A checkpoint is a predetermined point on the surface of the Earth used as a means of controlling movement, a registration target for fire adjustment or reference for location (JP 1-02). Checkpoints are a means of control between the parent unit and the patrol. These locations are decided upon and coordinated before the patrol leaves, so that both the patrol members and parent unit will know the patrol's location when it reports in. The parent unit can follow the progress of the patrol without transmitting coordinates over the radio that the enemy could monitor.

Rally Points

A rally point is an easily identifiable point on the ground where units can reassemble and reorganize if they become dispersed (MCRP 5-12C). It should provide cover and concealment, be defensible for at least a short time, and be easily recognized and known to all patrol members. All rally points are considered and identified as tentative rally points until they are occupied, found to be suitable, and designated as rally points by the patrol leader.

Initial Rally Point

The initial rally point is within friendly lines where the patrol can rally if it becomes separated before departing the friendly area or before reaching the first en route rally point. It may be the assembly area where the patrol waits while the patrol leader contacts the last friendly position through which the patrol will pass. The location of the initial rally point must be coordinated with the forward unit commander.

En route Rally Points

En route rally points are between the initial rally point and the objective rally point; and from the objective rally point back to the point where the patrol reenters friendly lines. They are determined as the patrol passes through a likely area that is suitable for a rally point.

Objective Rally Point

The objective rally point is located nearest the objective where the patrol makes final preparations prior to approaching the objective. It also serves as a location where the patrol reassembles after completing actions on the objective. The objective rally point must be suitable to accommodate those activities accomplished prior to actions on the objective. This position must provide the patrol concealment from enemy observation and, if possible, cover from enemy fires. It may be located short of, to a flank or beyond the objective. It should be out of sight, sound, and small-arms range of the objective area. The patrol leader's reconnaissance of the objective is made from this position; it is the release point from which patrol elements and teams move to the objective to accomplish the mission.

Rally Point Selection

The patrol leader selects likely locations for tentative rally points during reconnaissance or map study. A tentative rally point must be confirmed and announced after examination proves suitability. A tentative initial rally point and a tentative objective rally point are always selected and identified in the patrol order. If necessary, the patrol leader selects additional rally points en route as suitable locations are reached. When the patrol reaches a danger area that cannot be bypassed, such as an open meadow or stream, the patrol leader selects a rally point on both the near and far side. If good locations are not available, the patrol leader designates the rally points in relation to the danger area. The patrol leader will say, for example, "50 meters this side of the trail" or "50 meters beyond the stream."

Rally Point Use

The initial rally point and en route rally points are designated to enable the patrol to reassemble if it is unavoidably separated or dispersed. Identifying features are pointed out. The patrol leader ensures that the information is passed to all patrol members. When crossing a danger area, a near side rally point and a far side rendezvous point are designated. These rally points should only be used when all other methods of retaining control of the patrol have failed. The success of the patrol may be jeopardized if it is dispersed and cannot rally expeditiously.

If the patrol has left the friendly area and becomes dispersed, patrol members return to the last designated rally point (the initial or an en route rally point) unless the patrol leader gives other instructions.

As previously noted, the patrol leader selects two rally points at the near and far sides of danger areas that cannot be bypassed. If the patrol becomes separated or dispersed at a danger area, and there has been no enemy contact, the patrol should reassemble at the rally point on the far side of it. If, however, the patrol is separated or dispersed at a danger area as a result of enemy contact, members who have already crossed the danger area assemble and reorganize at the rally point on the far side; members who have not crossed assemble and reorganize at the rally point on the near side. In the absence of the patrol leader and unless directed otherwise, the senior member at the rally point on the near side takes charge, attempts to move the rallied patrol members to the rally point on the far side, and rejoins the remainder of the patrol.

Rally Point Actions

The patrol leader plans the actions to be taken at rally points and instructs the patrol accordingly in the patrol order. Planned actions at the initial rally point and en route rally points must provide for the continuation of the patrol as long as there is a reasonable chance to accomplish the mission. Plans for actions at rally points should provide for—

- Recognition signals for assembly at rally points.
- Minimum number of members and maximum amount of waiting time required before the senior member at the rally point moves the rallied patrol members onward toward the objective or returns to friendly lines.
- Instructions for patrol members who find themselves alone at a rally point.

11004. PRECAUTIONS AT DANGER AREAS

A danger area is any place where the patrol is vulnerable to enemy observation or fire (open areas, roads, trails, and obstacles such as barbed wire, minefields, rivers and streams, and lakes). Any known or suspected enemy position the patrol must pass is also a danger area. The patrol leader plans for crossing each danger area and includes these plans in the order.

The patrol reconnoiters the near side of a danger area first, then the patrol leader sends scouts to reconnoiter the far side. Once the scouts report that the far side is clear of the enemy, the remainder of the patrol crosses the danger area. As each individual or group crosses the danger area, they are covered by those remaining and by those who have successfully crossed. Enemy obstacles are avoided since they are usually covered by fire.

In crossing a river, the near bank is reconnoitered first; then the patrol is positioned to cover the far bank. Scouts are sent across to the far bank. After the far bank has been reconnoitered and the scouts report that it is clear of the enemy, the patrol crosses as rapidly as possible. This may be done individually or in pairs. If crossing the river requires swimming, the patrol uses improvised rafts to float equipment, weapons, and ammunition across. (Refer to MCRP 3-02C, *Water Survival Handbook.)*

A road or trail is crossed at or near a bend or where the road is narrow. Observation is restricted and, if the enemy is present, the patrol is exposed as short a time as possible. The near side is reconnoitered first, then scouts are sent across to reconnoiter the far side. This includes reconnoitering the tentative rally point on the far side. Once the scouts report "all clear," the remainder of the patrol crosses rapidly and quietly.

If the patrol must pass close to an enemy position, it takes advantage of battlefield noises to cover the sounds of movement. If supporting fires are available, the patrol leader can call for them to divert the enemy's attention as the patrol passes.

11005. HIDE

When a patrol is required to halt for an extended period in an area not protected by friendly troops, the patrol moves into a location which, by the nature of the surrounding terrain, provides passive security from enemy detection. Such an assembly area is termed a *hide*. To establishment a hide—

- Cease all movement during daylight hours to avoid detection.
- Hide the patrol for an extended period while the patrol leader conducts a detailed reconnaissance of the objective area.
- Rest and reorganize after extended movement.
- Reorganize after a patrol has infiltrated the enemy area in small groups.

The patrol leader's plan must include tentative hide locations when the patrol's mission dictates an extended halt within enemy areas. These tentative locations must be confirmed by actual ground reconnaissance prior to occupation by the patrol. The plan for a hide includes both passive and active security measures.

Passive security measures are—

- Avoid built-up areas.
- Select an area remote from all human habitation.
- Avoid known or suspected enemy positions.
- Avoid ridge lines, topographic crests, valleys, lakes, and streams.
- Avoid roads and trails.
- Avoid open woods and clearings.
- Select areas offering dense vegetation, preferably bushes and trees that spread out close to the ground.

Active security measures—

- Establish security covering all likely avenues of approach into the site.
- Establish communications (wire, radio, signal, runner) with posted security to provide early warning of enemy approach.
- Select an alternate area for occupation if the original hide is compromised or found unsuitable.
- Plan for withdrawal in the event of discovery.
- Establish an alert plan with a certain percent of the personnel awake at all times.
- Organize the elements of the patrol so necessary activities can take place with a minimum amount of movement.

The size of the area physically occupied by a patrol in a hide and the number of security posts required are governed by the terrain, quantity and quality of cover and concealment, and size of the patrol.

If the situation permits, a hide can also be used as the final preparation position and/or objective rally point.

11006. IMMEDIATE ACTIONS UPON ENEMY CONTACT

A patrol may make contact with the enemy at any time. Contact may be through observation, a meeting engagement or ambush. Contact may be visual, in which the patrol sights the enemy but is not itself detected. When this is the case, the patrol leader can decide whether to make or avoid physical contact, basing his decision on the patrol's assigned mission and capability to successfully engage the enemy unit.

When a patrol's assigned mission prohibits physical contact (except that necessary to accomplish the mission), its actions are defensive in nature. Physical contact, if unavoidable, is broken as quickly as possible and the patrol, if still capable, continues its mission.

When a patrol's assigned mission permits or requires it to seek or exploit opportunities for contact (as in the case of a combat patrol), its actions are offensive in nature, immediate, and positive.

In patrolling, contacts (visual or physical) are often unexpected at very close ranges, and short in duration. Effective enemy fire often provides leaders little or no time to fully evaluate situations and issue orders. In these situations, immediate action provides a means for swiftly initiating positive offensive or defensive action, as appropriate.

Two types of physical contact with the enemy are meeting engagement and ambush. Meeting engagement is a combat action that occurs when a moving force, incompletely deployed for battle, engages an enemy at an unexpected time and place. It is an accidental meeting where neither the enemy nor the patrol expect contact and are not specifically prepared to deal with it. An ambush is a surprise attack from a concealed position.

Immediate Actions

Immediate actions are designed to provide swift and positive small unit reaction to visual or physical contact with the enemy. They are simple courses of action in which all Marines are well trained. Minimal signals or commands are required and they are developed as needed for the combat situation. The signals can, in many cases, be initiated by any member of the unit. It is not feasible to attempt to design an immediate action drill to cover every possible situation. It is better to know the immediate action drill for each of a limited number of situations that may occur during a patrol. Arm-and-hand signals associated with immediate actions—such as FREEZE, ENEMY IN SIGHT, and HASTY AMBUSH RIGHT or LEFT—are contained in FMFM 6-5 (proposed MCWP 3-11.2).

Immediate Halt

When the patrol detects the enemy but is not itself detected, the situation requires the immediate, in-place halt of the patrol. The first member visually detecting the enemy gives the silent signal for FREEZE. Every member halts in place, weapon at the ready, and remains absolutely motionless and quiet until further signals or orders are given.

Air Observation and/or Attack

These actions are designed to reduce the danger of detection from aircraft and casualties from air attack.

When an enemy or unidentified aircraft that may detect the patrol is heard or observed, the appropriate immediate action drill is FREEZE. The first member hearing or sighting an aircraft that may be a threat signals FREEZE. Every member halts in place until the patrol leader identifies the aircraft and gives further signals or orders. Members of the patrol must not look up at the aircraft as sunlight can reflect off their faces even when camouflaged.

When an aircraft detects a patrol and makes a low level attack, the immediate action drill air attack is used. The first member sighting an attacking aircraft shouts, "AIRCRAFT," followed by the direction of the incoming attack: FRONT, LEFT, REAR or RIGHT. The patrol moves quickly into line formation, well spread out, at right angles to the aircraft's direction of travel. As each member comes on line, the member hits the ground, using available cover, then positions the body perpendicular to the aircraft's direction of travel, to present the shallowest target possible (see fig. 11-2 on page 11-8). Between attacks (if the aircraft returns or if more than one aircraft attacks), patrol members seek better cover. Attacking aircraft are fired upon only on command of the patrol leader.

Meeting Engagement

Hasty Ambush. This immediate action is used to avoid contact and to prepare to initiate an unplanned ambush on the enemy. It may often be a subsequent action after the command freeze. When the signal HASTY AMBUSH is given (by the point member, patrol leader or another authorized patrol member), the entire patrol moves quickly to the right or left of the line of movement, as indicated by the signal, and takes up the best available concealed firing positions (see fig. 11-3 on page 11-9). The patrol leader initiates the ambush by opening fire and shouting, "FIRE"; thus ensuring initiation of the ambush if the weapon misfires. If the patrol is detected before this, the first member aware of detection initiates the ambush by firing and shouting. The patrol leader may decide not to initiate the ambush in order to avoid contact unless the patrol is detected. When used as an offensive

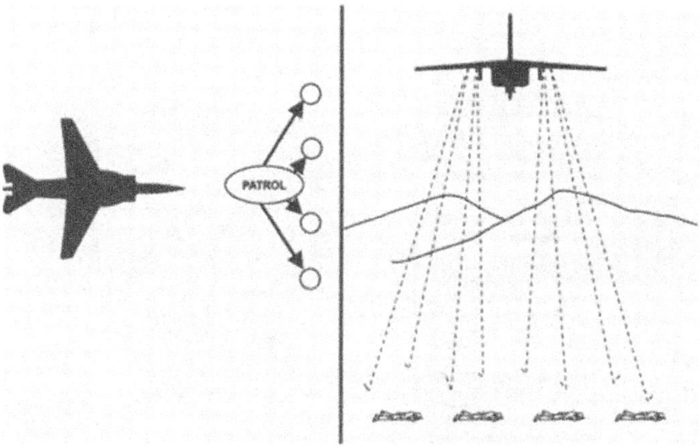

Figure 11-2. Immediate Action, Air Attack.

measure, the enemy is allowed to advance until he is in the most vulnerable position before the ambush is initiated. An alternate means for initiating the ambush is to designate an individual (for example, point or last member) to open fire when a certain portion of the enemy unit reaches or passes that member.

Immediate Assault. This immediate action drill is used defensively to make and quickly break undesired but unavoidable contact (including ambush) and offensively to decisively engage the enemy (including ambush). When used in a meeting engagement, members nearest the enemy open fire and shout, "CONTACT," followed by the direction of the incoming attack: FRONT, LEFT, REAR or RIGHT. The patrol moves swiftly into line formation and assaults (see fig. 11-4).

Defensive Measures

When used defensively, the assault is stopped if the enemy withdraws and contact is broken quickly. If the enemy stands fast, the assault is carried through the enemy positions and movement is continued until contact is broken.

Offensive Measures

When used offensively, the enemy is decisively engaged. Escapees are pursued and destroyed until orders to break contact are given by the patrol leader. If the patrol is fired upon from beyond 50 meters, the patrol must break contact as quickly as possible and continue the mission. If it engages the enemy any longer than necessary to break contact, it may put the mission in jeopardy.

Fire and Maneuver

Fire and maneuver is one means to break contact. One portion of the patrol returns the enemy fire while another portion moves by bounds away from the enemy. Each portion of the patrol covers the other by fire until contact is broken by all.

Clock System

The clock system is another means to break contact. Twelve o'clock is the direction of movement of the patrol. The patrol leader shouts a direction and a distance. For example: "TEN O'CLOCK-TWO HUNDRED," means the patrol should move in the direction of ten o'clock for 200 meters. Patrol members keep their same relative positions as they move so the original formation is not disrupted. Subordinate leaders must be alert to ensure that the members of their elements and teams receive the correct order and move as directed.

Counter Ambush

When a patrol is ambushed, the immediate action drill used depends on whether the ambush is a near ambush

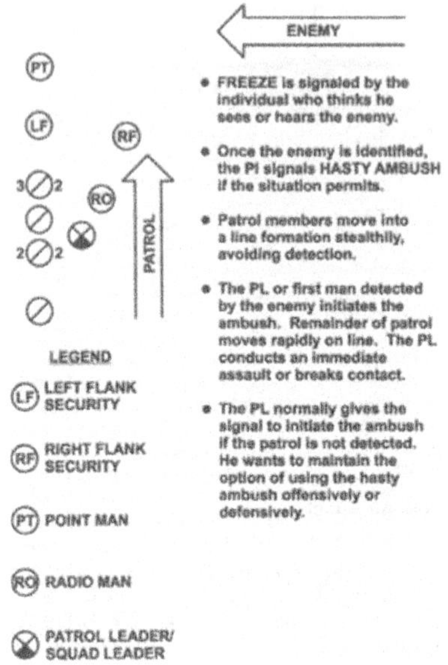

Figure 11-3. Immediate Action, Hasty Ambush.

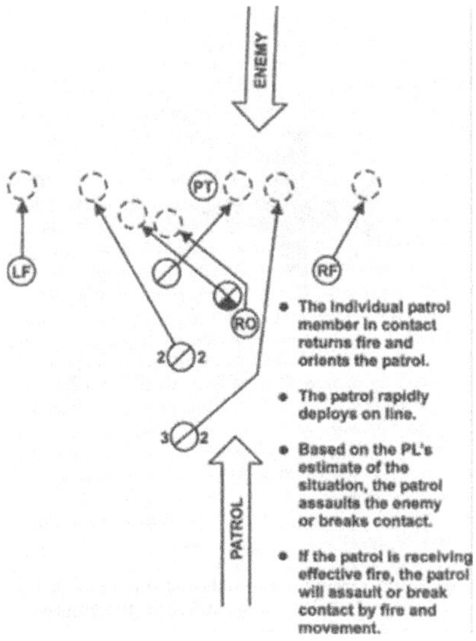

Figure 11-4. Immediate Action, Immediate Assault.

(the enemy is within 50 meters of the patrol) or a far ambush (the enemy is beyond 50 meters of the patrol). Fifty meters is considered the limit from which the ambushed patrol can effectively launch an assault against the enemy.

Near Ambush. In a near ambush, the killing zone is under very heavy, highly concentrated, close range fires. There is little time or space for members to maneu-ver or seek cover. The longer they remain in the killing zone, the greater the chance they will become casualties. Therefore, if members in the killing zone are attacked by a near ambush, they immediately assault without order or signal directly into the ambush position, occupy it, and continue the assault or break contact, as directed. This action moves them out of the killing zone, prevents other elements of the ambush from firing on them without firing on their own members, and provides positions from which other actions may be taken (see fig. 11-5 on page 11-10). Members not in the killing zone maneuver against the enemy as directed. The assault continues until all patrol members are outside of the killing zone.

Far Ambush. In a far ambush, the killing zone is also under very heavy, highly concentrated fires, but from a greater range. This greater range provides members in the killing zone maneuver space and some opportunity to seek cover at a lesser risk of becoming a casualty. If attacked by a far ambush, members in the killing zone, without order or signal, immediately return fire, take the best available positions, and continue firing until directed otherwise. Members not in the killing zone maneuver against the ambush force, as directed (see fig. 11-6 on page 11-10). The assault is continued against the enemy or until the order to break contact is given.

In each situation, the success of the counter ambush employed depends on the members being well trained in quickly recognizing the distance from which an ambush is initiated and well rehearsed in the proper reaction.

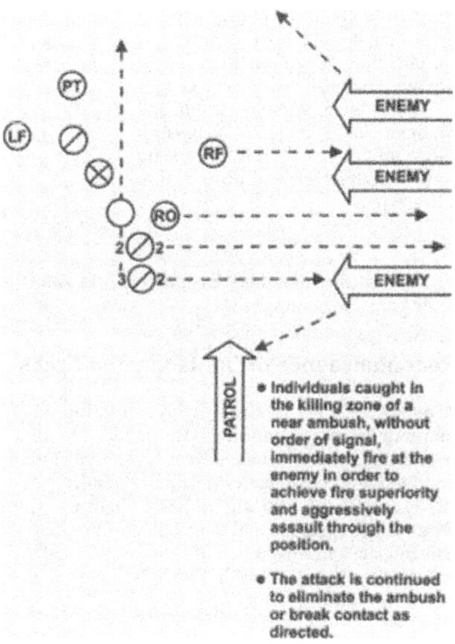

Figure 11-5. Near Ambush.

11007. PATROL LEADER'S ACTION IN A DEVELOPING SITUATION

While good patrolling depends on good planning, the patrol leader's plan must be flexible. Every combat situation develops differently than expected, and the patrol leader must be ready to quickly adapt to the situation as it develops. One of the most difficult tasks in battle is to recognize the correct moment for making a decision. Generally, it is more difficult to determine the moment for making a decision than it is to formulate the decision itself. When the situation demands, decisions must be made promptly without waiting for more information. In a developing situation, the patrol leader should use the following questions as a guide for battlefield decisionmaking:

- How has the situation changed?
- How does the change affect mission accomplishment and the immediate superior's mission?
- Must a decision be made now?
- What are the options?
- Which option best serves the mission and the unit as a whole?
- Which option offers the greatest chance of success?

11008. RETURN FROM OBJECTIVE AREA

After performing actions in the objective area, the patrol reassembles at the objective rally point. This phase of the patrol is perhaps the most difficult and dangerous. Patrol members are experiencing fatigue, emotional letdown, and wounds; they may be low on water and ammunition. Above all, the enemy was likely alerted if not in pursuit. At this point, the patrol leader must move the patrol rapidly but carefully and maintain patrol security at a high level. In returning to friendly lines, the patrol neither uses nor travels near the same route used to get to the objective area. The enemy may have that route covered. For information on reentry of friendly lines, refer to paragraph 11001.

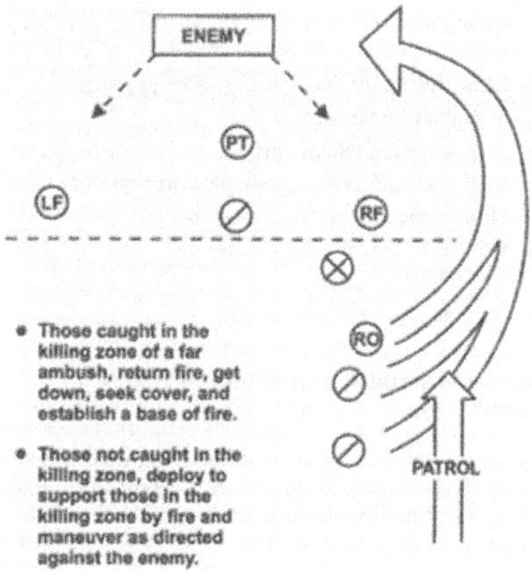

Figure 11-6. Far Ambush.

CHAPTER 12. RECONNAISSANCE PATROLS

The commander needs accurate, timely information about the enemy and the terrain the enemy controls to assist in making tactical decisions. Reconnaissance patrols are one of the most reliable means for obtaining this information. These patrols engage the enemy only when necessary to accomplish their mission or for protection. They generally avoid combat, accomplish the mission by stealth, and do not maintain contact with the unit that sent them out. A reconnaissance patrol is capable of carrying the search for information into the area occupied by enemy forces—usually beyond the range of vision of friendly ground observation posts—and examining objects and events at close range.

12001. GENERAL MISSIONS

General missions for reconnaissance patrols include gathering information about location and characteristics of friendly or hostile positions and installations, terrain (routes, stream crossings), and obstacles.

12002. SPECIFIC MISSIONS

Locate the Enemy

Reconnaissance patrols try to determine the—

- Location of enemy forces, installations, and equipment.
- Identification of enemy units and equipment.
- Strength of enemy forces.
- Disposition of enemy forces.
- Movement of enemy personnel and equipment.
- New or special types of weapons.
- Presence of mechanized units.
- Unusual enemy activity.
- Presence of NBC equipment.

Reconnaissance of Enemy Wire Obstacles

A patrol with the mission of investigating enemy-emplaced wire obstacles employs a formation providing for all-around security and takes precautions against being observed by the enemy. The patrol leader and one Marine inspect each gap and establish its location by means of compass bearings to prominent objects in the rear of enemy or friendly positions or through the use of global positioning satellite receivers.

Reconnaissance of Contaminated Areas

A patrol with the mission of investigating a contaminated area reconnoiters and marks the area's boundaries. Patrol members wear protective clothing and field protective masks. The patrol leader reports the extent of the area, the ttype of agent used, the terrain and vegetation and the method of marking the contaminated area. A sketch of the contaminated area, prepared by the patrol leader, should be included in the patrol report.

Reconnaissance of Enemy Minefields

Patrols assigned to reconnoiter enemy minefields are often composed of specially trained personnel. The type of mine is identified and the possibility of bypassing the area determined.

Reconnaissance of Terrain

The reconnaissance patrol is particularly suited for gathering information about the terrain within their area of operations. (Refer to MCRP 2-15.3B, *Reconnaissance Reports Guide*.) Commanders must know the location and condition of bridges, streams, and roads in order to make proper plans for the movement of troops. The following information should be obtained by a reconnaissance patrol assigned such missions:

- Bridges:
  - Maximum load capacity.
  - Material used in construction (wood, stone, concrete or steel).

- Material used for abutments (dirt, wood, masonry, concrete).
- Type and number of supporting members of the bridge.
- Condition and dimensions of bridge flooring.
- Presence or absence of mines and demolition charges on the bridge and at entrances/exits.
- Terrain crossed (ravine, stream, road).
- Primary use (rail, auto, foot).
- Location and type of an alternate crossing if bridge is unusable.

- Streams and fords:
  - Width and depth of stream.
  - Composition of the bottom (mud, sand, gravel, rock).
  - Speed of current in miles per hour.
  - Composition and gradient of banks.
  - Surrounding terrain and vegetation cover.
- Roads:
  - Width.
  - Composition of surface.
  - Condition.
  - Road blocks.
  - Gradient/degrees of slopes.
  - Curves (location and length).
  - Pitch and culvert locations.
  - Primary use and frequency of use.

12003. TYPES OF RECONNAISSANCE

Route

Route reconnaissance is a directed effort to obtain detailed information of a specified route and all terrain from which the enemy could influence movement along that route. (Refer to MCRP 5-12C.) It is also described as a form of reconnaissance focused along a specific line of communications—such as a road, railway, or waterway—to provide new or updated information on route conditions and activities along the route. (Refer to MCRP 5-12A, *Operational Terms and Graphics*.)

A route reconnaissance normally precedes the movement of forces. It provides detailed information about a specific route and the surrounding terrain that could be used to influence movement along that route. Considerations include trafficablility, danger areas, critical points, vehicle weight and size limitations, and locations for friendly obstacle emplacement.

The objective area is normally defined by a line of departure, a route, and a limit of advance.

Area

An area reconnaissance is a directed effort to obtain detailed information concerning the terrain or enemy activity within a prescribed area such as a town, ridge line woods or other features critical to operations. (Refer to MCRP 5-12C.) An area reconnaissance could also be made of a single point, such as a bridge or installation. (Refer to MCRP 5-12A.)

Emphasis is placed on reaching the area without being detected. Enemy situations encountered en route are developed only enough to allow the reconnoitering unit to report and bypass.

Zone

A zone reconnaissance is a directed effort to obtain detailed information concerning all routes, obstacles (to include chemical or radiological contamination), terrain, and enemy forces within a zone defined by boundaries. A zone reconnaissance is normally assigned when the enemy situation is vague or when information concerning cross-country trafficability is desired. (Refer to MCRP 5-12C.)

The commander specifies specific routes or areas of interest within the zone. The zone to be reconnoitered usually is defined by a line of departure, lateral boundaries, and a limit of advance.

12004. TASK ORGANIZATION

A reconnaissance patrol is organized around the current structure of the Marine rifle squad with one or more of the squad's fire teams assigned as the reconnaissance element to reconnoiter or maintain surveillance over the objective. At least one fire team acts as a security element whose functions are to—

- Secure the objective rally point (see para. 11003).
- Give early warning of enemy approach.
- Protect the reconnaissance unit.

Reinforcing the squad is based on METT-T. If two squads are required, then the mission should be assigned to a platoon.

A company may send a reconnaissance patrol to specifically reconnoiter a given area (a riverbed or a bridge), or to maintain surveillance on a particular streambed for the next four nights. Reconnaissance patrols perform three basic functions en route to and from the objective:

- Provide control.
- Provide security while moving and in the objective area.
- Conduct reconnaissance or surveillance.

Depending on the size of the reconnaissance patrol, an element must be tasked to reconnoiter the area or zone. In the case of a surveillance mission, two teams are recommended to share the same task (so one could lay up and rest, while the other lays low and hides to maintain strict silence and no movement while surveillance is being maintained on the objective). Whatever the case, each unit that has a surveillance mission or reconnaissance mission must be thoroughly briefed as to what essential elements of information are to be collected as well as their location in the objective area.

The patrol should be organized with one or two fire teams to actually conduct the reconnaissance mission and the remaining fire team to provide security. Each fire team should be prepared to assume either mission. A small area reconnaissance patrol needs only one fire team for the assigned mission. A patrol with a wider area reconnaissance mission should use two fire teams to physically conduct the mission and one fire team for cover and/or security.

The security element for a reconnaissance patrol should be organized to cover the likely avenue of approach into the objective area, to protect the units conducting the reconnaissance, and to cover the objective rally point.

The variations of the special organization for reconnaissance patrols illustrate the patrol leader's flexibility in organizing the patrol to meet mission requirements. The patrol leader is not limited to the variations reflected herein but can choose any special organization as long as that task organization accommodates the requirement for command and control, reconnaissance, and security.

12005. SIZE OF RECONNAISSANCE PATROLS

A reconnaissance patrol should be kept to the minimum number of personnel required to accomplish the mission. A mission requiring a patrol to remain away from its unit for a considerable period of time, or one requiring a patrol to send back information by messenger, increases the size of the patrol. Reconnaissance patrols seldom exceed a squad in strength. Unit integrity should be preserved whenever possible. Intelligence personnel, interpreters, and other specialists, such as radio operators or engineers, are assigned to a patrol if the particular mission requires.

12006. RECONNAISSANCE EQUIPMENT

Patrol members are armed and equipped as necessary for accomplishing the mission. The automatic rifle in each fire team provides a degree of sustained firepower in case of enemy contact. The patrol should have at least two: pairs of binoculars, pairs of wire cutters, maps, compasses, and watches. Night observation equipment may be used. Pencils and small notebooks are carried so notes and sketches can be made. A message book with message blanks and overlay paper is mandatory.

12007. RECONNAISSANCE PATROL ACTIONS AT THE OBJECTIVE AREA

Route

The patrol leader halts and conceals the patrol near the objective area in the final preparation position. The patrol leader then conducts leader's reconnaissance to confirm the plan for positioning the security teams and employing units assigned to the reconnaissance mission. The patrol leader returns to the patrol and positions the security to provide early warning of enemy approach and secure the objective rally point. The reconnaissance unit(s) then reconnoiters the objective area (route). The reconnaissance unit may move to several positions, along or adjacent to the specific route, in order to conduct a thorough reconnaissance. After completing the reconnaissance, each reconnaissance team moves to the objective rally point and reports to the patrol leader. The patrol then returns to friendly lines and the leader makes a full report.

Area

The patrol leader halts and conceals the patrol near the objective area in the final preparation position. The patrol leader then conducts leader's reconnaissance to pinpoint the objective and confirm the plan for positioning the security teams and employing units assigned the reconnaissance mission. The patrol leader returns to the patrol and positions security to provide early warning of enemy approach and secure the objective rally point. The reconnaissance unit(s) then reconnoiters the objective area. The reconnaissance unit may move to several positions, perhaps making a circle around the objective area, in order to conduct a thorough reconnaissance. When the reconnaissance is completed, the patrol leader assembles the patrol and tells members what has been observed and heard. Other patrol members contribute observations. The patrol then returns to friendly lines and the patrol leader makes a full report.

Zone

The patrol leader halts the patrol at the final preparation position, confirms the plan, and conducts leader's reconnaissance. The patrol leader positions the security team and sends out the reconnaissance team. When the entire patrol is used to reconnoiter the zone, it provides its own security. After completing the reconnaissance, each reconnaissance team moves to the objective rally point and reports to the patrol leader. The patrol then returns to friendly lines.

CHAPTER 13. COMBAT PATROLS

Combat patrols are assigned missions that usually include engaging the enemy. They are fighting patrols. Every combat patrol has a secondary mission: gaining information about the enemy and terrain. Combat patrols are employed in both offensive and defensive combat operations and they assist the parent unit in accomplishing its mission by inflicting damage on the enemy; establishing and/or maintaining contact with friendly and enemy forces; denying the enemy access to key terrain; and probing enemy positions to determine the nature and extent of enemy presence.

13001. TASK ORGANIZATION

A combat patrol is organized around the current structure of the Marine rifle platoon. A combat patrol leader should use the unit's normal organization (fire team, squad, and platoon) in assigning functions, patrol missions, and chain of command. Combat patrols must be able to perform the following four basic functions en route to and from the objective:

- Provide control.
- Provide security.
- Provide support by fire.
- Attack or assault the objective.

A rifle platoon could task-organize as a combat patrol as follows:

- Platoon headquarters (patrol headquarters).
- First squad (security).
- Second squad (support).
- Third squad (assault).

Every combat patrol must—

- Provide a control mechanism in the form of a headquarters.
- Designate a unit (a fire team or squad) to provide security while moving en route to the objective or while at the objective. At the objective area, this unit isolates the objective area, secures the objective rally point, and covers the withdrawal.
- Designate a unit to act as support. This unit provides the base of fire in the attack or covers withdrawals or advances.
- Designate a unit(s) to conduct the attack or assault. This unit(s) engages the enemy at the objective area by fire and maneuver or movement. It also operates immediately in the objective area (searching, demolition, prisoners of war, etc.). Paragraph 9001 outlines the general organization of combat patrols. As in the case with reconnaissance patrols, the task organization of a combat patrol depends on the specific mission assigned. If any special requirements are generated because of the specific mission, the patrol is task-organized to fit the needs of the mission.

13002. EQUIPMENT

Combat patrols are armed and equipped as necessary for accomplishing the mission. In addition to binoculars, wire cutters, compasses, and other equipment generally common to all patrols, it usually carries a high proportion of automatic weapons and grenades. Communications with higher headquarters is important as success of the mission may depend on being able to call for supporting fires. Also, internal radio communications with the units and teams may be useful. However, the patrol must not be so overburdened with equipment as to impede movement or mission accomplishment.

13003. RAID PATROLS

A raid is a surprise attack on an enemy force or installation with the attacking force withdrawing after accomplishing its mission. Raids destroy or capture enemy personnel or equipment, destroy installations, or free friendly personnel who have been captured by the enemy. Patrolling techniques are used in planning and when moveing to and from the objective. (Refer to MCWP 3-41.2, *Raids*.) Surprise, firepower, and violence of action are the keys to a successful raid. Patrols achieve surprise by attacking—

- When the enemy is least prepared (e.g., during periods of poor visibility such as darkness, rain, fog, or snow).
- From an unexpected direction. (This might be accomplished by approaching through a swamp or other seemingly impassable terrain.)
- With concentration of firepower at critical points within the objective.

Planning

A successful raid requires detailed planning. The leader of a combat patrol engaged in raiding must anticipate probable situations and decide upon definite courses of action to meet them. Rehearsals are imperative.

A raid patrol conducts such missions as destroying an enemy outpost or seizing prisoners from an observation post or lightly defended position.

While preparing for the mission, the patrol leader requests fire support required for the accomplishment of the mission. If practical, artillery and mortars should be employed to isolate the objective to prevent movement of enemy reinforcements into the area.

Execution

The leader's plan must be detailed and complete. All of the considerations outlined in chapter 11 must be covered. Patrol formations must provide for ease of control and all-around security while moving to and from the objective area and provide for rapid and coordinated deployment of the various units once the objective area is reached. The leader's plan usually includes the encirclement of the hostile position—either physically or by fire—in order to isolate it during the assault.

The final simultaneous assault against the objective develops when enemy defensive fires at the objective are suppressed by either friendly fire superiority or surprise. The assault is covered by the fire of the unit assigned the function of support by fire.

Grenades, SMAWs, and demolitions are most effective for clearing bunkers.

Security units are posted to isolate the objective. The patrol leader signals them when the withdrawal begins. As a minimum, security is on each flank and to the rear (at the objective rally point).

Actions at the Objective

The patrol leader halts the patrol near the objective at the final preparation position. Security is established and the leader's reconnaissance is made with appropriate subordinate leaders. When the leaders return to the patrol, they confirm previous plans or announce any changes. Movements are arranged so all units reach their positions simultaneously. This improves the patrol's capability for decisive action, if prematurely detected by the enemy.

The teams of the security element move to positions to secure the objective rally point, give early warning of enemy approach, block avenues of approach into—and prevent enemy escape from—the objective area. As the assault element moves into position, the security element informs the patrol leader of all enemy activity, firing only if detected or on the patrol leader's order. Once the assault element commences action, the security element prevents enemy entry into or escape from the objective area. The security element covers the withdrawal of the assault element (and support element, if employed) to the objective rally point, withdrawing only upon order or prearranged signal.

As the assault element approaches the objective, it deploys early enough to permit immediate assault if detected by the enemy. Each team uses stealth while moving into proper position. On command, or if one or more of the assault element is detected and fired upon by the enemy, the support element opens fire to neutralize the objective, then ceases or shifts fire according to prearranged plans and signals. As supporting fires cease or shift, the assault element assaults the objective. Demolition, search, and other teams are protected by the assault element while they work. On order, the assault element withdraws to the objective rally point.

If a support element is employed, its leader deploys teams to provide fire support for the assault element. Each member of the support element must know the scheme of maneuver to be used by the assault element, specific targets or areas to be neutralized by fire, and the signals that will be employed to commence, shift, and cease fires. The support element withdraws on order of the patrol leader. At the objective rally point, the patrol leader quickly reorganizes the patrol and begins the return movement to friendly lines.

13004. CONTACT PATROLS

General

Contact patrols establish and/or maintain contact to the front, flanks or rear by—

- Establishing contact with an enemy force when the definite location of the force is unknown.
- Maintaining contact with enemy forces through direct and/or indirect fires, or observation.
- Avoiding decisive engagement with the enemy.

Task Organization and Equipment

Task organization and equipment depend on the known enemy situation and anticipated enemy contact. A patrol sent out to establish contact with an enemy force is organized, armed, and equipped to overcome resistance of light screening forces in order to gain contact with the main enemy force. It is not organized and equipped to engage the main enemy forces in combat. Communication is paramount; radios must be reliable over the entire distance covered.

Actions at the Objective

The patrol leader selects a series of objectives. Once an objective is reached, the patrol leader initiates a planned set of actions in order to establish and maintain contact with the enemy. These plans and actions are guided by the missions to establish or maintain contact—not to engage in decisive combat. Contact with the enemy is maintained for the purposes of surveillance, applying pressure, and preventing seizure of the initiative. If the contact patrol becomes decisively engaged with the enemy, many of the tasks originally assigned to the patrol cannot be accomplished, since the enemy has seized the initiative and friendly forces are not forced to react.

13005. AMBUSH PATROLS

General

An ambush is a surprise attack from a concealed position upon a moving or temporarily halted target. It is one of the oldest and most effective types of military actions. Ambush patrols conduct ambushes of enemy patrols, resupply columns, and convoys. The intent of an ambush is to place the enemy in a dilemma where staying in the kill zone or attempting to move out of it prove equally lethal. The ambush may include an assault to close with and decisively engage the enemy, or the attack may be by fire only.

Purpose of Ambushes

Ambushes are executed for the general purpose of reducing the enemy's over-all combat effectiveness and for the specific purpose of destroying its units. The cumulative effect of many small ambushes on enemy units lowers enemy troop morale and harasses the enemy force as a whole.

Destruction is the primary purpose of an ambush because loss of men killed or captured, and loss of equipment and supplies destroyed or captured, reduces the overall combat effectiveness of the enemy.

Harassment, though less apparent than physical damage, is a secondary purpose of ambushes. Frequent ambushes force the enemy to divert men from other missions to guard convoys, troop movements, and carrying parties. When enemy patrols fail to accomplish their mission because they are ambushed, the enemy is deprived of valuable information. A series of successful ambushes causes the enemy to be less aggressive and more defensive minded. His men become apprehensive, overly cautious, reluctant to go on patrols, seek to avoid night operations, are more subject to confusion and panic if ambushed, and in general, decline in effectiveness.

Classification of Ambushes

A *deliberate ambush* is one in which prior information about the enemy permits detailed planning before the patrol departs for the ambush site. Information needed to plan a deliberate ambush includes the size, composition, and organization of the force to be ambushed; how the force operates; and the time it will pass certain points or areas. A deliberate ambush may be planned for such targets as—

- Any force if sufficient prior information is known.
- Enemy patrols that establish patterns by frequent use of the same routes or habitually depart and re-enter their own areas at the same point.
- Logistic columns.
- Troop movements.

An *ambush of opportunity* is conducted when available information does not permit detailed planning before the patrol departs. This is the type of ambush that an infantry unit normally conducts. An ambush of opportunity should not be confused with a hasty ambush. An ambush of opportunity is a planned ambush; a hasty ambush is an immediate action. In planning for an ambush of opportunity, the patrol must be prepared to execute any of several courses of action based on the types of targets that may be ambushed and must rehearse prior to departure. The course of action taken is determined when the opportunity for ambush arises.

The patrol leader may be directed to reconnoiter an area for a suitable ambush site, set up at the site selected, and execute an ambush against the first profitable target that appears.

The patrol may depart just after dark, move to a specific point, observe until a designated time, ambush the first profitable target after that time, and return before daylight.

A *hasty ambush* is an immediate action where the patrol makes visual contact with an enemy force and has time to establish an ambush without being detected. The actions for a hasty ambush must be well rehearsed and accomplished through the use of hand and arm signals given from the patrol leader.

Types of Ambushes

There are two types of ambushes: point and area. The *point ambush* is one where forces are deployed to attack along a single killing zone. The *area ambush* is one where forces are deployed as multiple related point ambushes.

A point ambush, whether independent or part of an area ambush, is positioned along the enemy's expected route of approach. Formation of the forces conducting the ambush is important because, to a great extent, the formation determines whether a point ambush is able to deliver the heavy volume of highly concentrated fire necessary to isolate, trap, and destroy the enemy.

The ambush formation to be used is determined by careful consideration of possible formations and the advantages and disadvantages of each in relation to—

- Terrain.
- Visibility.
- Forces.
- Weapons and equipment.
- Ease or difficulty of control.
- Target to be attacked.
- Combat situation.

For a detailed discussion of ambush formation, see appendix D.

Ambush Operation Terms

The *ambush site* is the location where an ambush is established.

The *killing zone* is that portion of an ambush site where fires are concentrated to trap, isolate, and destroy the target. On little-traveled roads, an obstacle placed in a defile, in the woods, on a bridge or on a steep upgrade can be used effectively to force vehicles to halt, and thus render the occupants vulnerable to attack. Antitank mines may be emplaced and the occupants of the wrecked vehicle killed or captured while still dazed by the explosion.

A *near ambush* is a point ambush where the attacking force is located within reasonable assaulting distance of the killing zone (50 meters is a guide). A near ambush is most often conducted in close terrain, such as a jungle or heavy woods.

A *far ambush* is a point ambush where the attack force is located beyond reasonable assaulting distance of the killing zone (beyond 50 meters is a guide). A far ambush may be more appropriate in open terrain offering good fields of fire or when the target will be attacked by fire only.

Factors for a Successful Ambush Patrol

There are many factors that give the ambush its best chance of success. The ideal situation would be to position the ambush on favorable terrain and have detailed planning completed beforehand.

The patrol leader of an ambush looks for the most favorable terrain in which the enemy is canalized between two obstacles with limited opportunity to attack or escape. Suitable areas include defiles, small clearings, bends in trails, and steep grades. Dense undergrowth adjacent to the ambush site permits observation from concealed positions. The ambush patrol should have maximum cover and concealment, not only for the firing positions, but for the routes of withdrawal. The enemy should be in an area offering as little protection from fire as possible. Favorable fields of fire include stretches of road, trail or open ground of at least 100 meters for machine guns and 15 meters for rifle fire and grenades. The ambush site can be improved by constructing obstacles—such as felled trees, wire, land mines, or booby traps—to impede the enemy.

Planning

A deliberate ambush or an ambush of opportunity requires thorough planning.

A deliberate ambush plan is based on extensive knowledge of the enemy and terrain, and is planned and rehearsed in great detail. A physical reconnaissance of the ambush site is made during the preparation phase and information gained is incorporated into the plan. All likely immediate actions of the enemy when ambushed are examined. Planned counteractions are developed and rehearsed.

In planning an ambush of opportunity, any available information on the enemy and terrain is used. A tentative plan for the ambush that incorporates all anticipated actions is developed and rehearsed. However, the bulk of planning is done concurrently during the patrol leader's reconnaissance of the prospective ambush site. In a rapidly developing situation, hasty ambush immediate action is employed.

The route and ambush site considerations apply to both deliberate ambushes and ambushes of opportunity. A primary route that allows the patrol to enter the ambush site from the rear is planned. Entering the prospective killing zone is avoided. If the killing zone must be entered to place mines or explosives, care is taken to remove any tracks and signs that might alert the enemy and compromise the ambush. If mines or explosives are to be placed on the far side of the ambush site, or if the appearance of the site from the enemy's viewpoint is to be checked, a wide detour is made around the killing zone. Care is taken to remove any tracks that might reveal the ambush. An alternate route from the ambush site to the objective rally point, as in other patrols, is planned.

Maps and aerial photographs are used to carefully analyze the terrain. When possible, an on-the-ground reconnaissance of the ambush site is made prior to occupation. Obvious ambush sites are avoided as the element of surprise is even more difficult to achieve in these areas. An ambush site must provide for—

- Favorable fields of fire.
- Occupation and preparation of concealed positions.
- Canalization of the target into the killing zone. (An ideal killing zone restricts the enemy on all sides, confining him to an area where he can be quickly and completely destroyed. Natural obstacles, such as cliffs, streams, embankments, or steep grades, are used whenever possible to force vehicles to slow down. Man-made obstacles, such as barbed wire, mines, and craters in the roads, are used to supplement natural obstacles.)
- Covered routes of withdrawal that enable the ambush force to break contact.
- Avoidance of enemy pursuit by fire.

Occupation of Ambush Site

The surrounding area is searched for enemy patrols prior to occupation of the ambush site. Ambush formations are used to physically deploy the patrol in a manner to inflict maximum destruction upon the enemy and to provide maximum security to the patrol. Ambush formations are contained in appendix D.

Positions

The patrol is moved into the ambush site from the objective rally point. Security is positioned first to prevent surprise while the ambush is being established. Automatic weapons are then positioned so each can fire along the entire killing zone. If this is not possible, overlapping sectors of fire are provided to cover the entire killing zone. The patrol leader then selects his position, located so he can tell when to initiate the ambush. Riflemen and grenadiers are positioned and sectors of fire are assigned to cover any dead space left by the automatic weapons. The patrol leader sets a time by which positions are to be prepared. Patrol members clear fields of fire and prepare positions in that order, with attention to camouflage for both.

Suitable Objective Rally Point

An easily located objective rally point is selected and made known to all patrol members. The objective rally point is located far enough from the ambush site so that it will not be overrun if the enemy assaults the ambush. Routes of withdrawal to the objective rally point are reconnoitered. Situation permitting, each man walks the route he is to use and picks out checkpoints. When the ambush is to be executed at night, each man must be able to follow his route in the dark. After the ambush has been executed, and the search of the killing zone completed, the patrol is withdrawn quickly but quietly, on signal, to the objective rally point where it reorganizes for the return march. If the ambush was not successful and the patrol is pursued, withdrawal may be by bounds. The last group may arm mines, previously placed along the withdrawal route, to further delay pursuit.

Local Security

Security must be maintained. Security elements do not usually participate in the initial attack, but protect the rear and flanks, and cover the withdrawal.

Patience

The Marines of the ambush force must control themselves so that the ambush is not compromised. Patience and self-discipline are exercised by remaining still and quiet while waiting for the target to appear, particularly if the patrol occupies the ambush site well ahead of the arrival of the enemy. Patience is necessary so as not to alert the enemy to the presence of the ambush.

Surprise

Surprise must be achieved, or the attack is not an ambush. If complete surprise cannot be achieved, it must be so nearly complete that the target is not aware of the ambush until too late for effective reaction. Surprise is achieved by careful planning, preparation, and execution so that targets are attacked when, where, and in a way for which they are least prepared.

Coordinated Fires

Properly timed and delivered fires contribute heavily to the achievement of surprise, as well as to destruction of the enemy. The lifting or shifting of fires must be equally precise; otherwise, the assault is delayed and the enemy has an opportunity to recover and react. All weapons, mines, and demolitions are positioned and all fires, including those of available artillery and mortars, are coordinated to achieve the following results:

- Isolation of the killing zone to prevent the enemy's escape or reinforcement.
- Surprise delivery of a large volume of highly concentrated fires into the killing zone.

Control

Close control of the patrol is maintained during movement to, occupation of, and withdrawal from the ambush site. This is best achieved through rehearsals and establishment and maintenance of good communications. When the enemy approaches, the temptation to open fire before the signal is given is resisted. The patrol leader must effectively control all elements of the ambush force. Control is most critical at the time the enemy approaches the killing zone. Control measures must provide for—

- Early warning of enemy approach.
- Fire control. Withhold fire until the enemy has moved into the killing zone, then open fire at the proper time.
- Initiation of appropriate action, if the ambush is prematurely detected. Individual patrol members must be prepared to react if detected by the enemy prior to the initiation of the ambush.
- Timely and orderly withdrawal of the ambush force from the ambush site and movement to the objective rally point.

It is important to remember that an ambush patrol should have four distinct signals: one to open fire (with an alternate signal to open fire to be used at the same time as the primary); a signal to cease fire or shift fire; a signal to assault or search the killing zone; and a signal to withdraw. The signal to open fire should meet two criteria: first, it should be the firing of a weapon that will kill the enemy; secondly, it should be a weapon reliable in any weather condition. A good primary signal is a Claymore mine, and an alternate signal would be a closed bolt weapon (M16A2). Open bolt weapons (M240G, M249) should not be relied upon to initiate an ambush.

Execution of an Ambush

The manner in which the patrol executes an ambush depends primarily on whether the ambush's purpose is harassment or destruction. To a lesser degree, the execution of the ambush is determined by whether the ambush is deliberate or an ambush of opportunity.

When the primary purpose is harassment, the patrol seals off the area with security teams to prevent enemy reinforcement and escape. Maximum damage is inflicted with demolitions and automatic weapons fire. The patrol delivers a very heavy volume of fire for a short time and withdraws quickly and quietly. The patrol avoids being seen by the enemy.

When the primary purpose of the patrol is destruction, the area is sealed off with security units. Maximum damage is inflicted with demolitions, antitank weapons, and automatic weapons fire from the support team or element. When these fires cease or shift, an assault is launched into the killing zone with heavy fire and violence to complete destruction. The assault unit provides security, while designated teams search and/or capture personnel and destroy vehicles and equipment. On the patrol leader's command, or by prearranged signal, all units withdraw to the objective rally point and move out quickly.

When the patrol's primary purpose is to obtain supplies or capture equipment, security units seal off the area. Demolitions and weapons are used to disable vehicles. The assault unit must use care to ensure its fire does not damage the desired supplies or equipment. Designated teams secure the desired items; other teams then destroy enemy vehicles and equipment.

The most successful ambush is one where the attacker is deployed and concealed in such a way that the enemy will unknowingly be surrounded by fire. The usual method is for the attackers to deploy themselves along a trail or route the enemy will travel. The enemy is permitted to pass by the center of the attacker's force so that the attack can be made from the front, flank, and/or rear. One or two men are posted well forward and to the rear along the route to prevent any enemy from escaping. All fires should be delivered simultaneously on a prearranged signal.

An effective method of luring the enemy is for an ambushing patrol to cut communication or electrical wire. The patrol then deploys and ambushes the line repair crew when it arrives. Since the line crew may be protected by riflemen, the attackers must be careful to engage the entire party.

Vehicles and foot personnel moving on well-established transportation routes can sometimes be captured by altering or moving directional signs so as to divert the enemy into an area where he can be more readily attacked. The attack can best be accomplished at an obstacle, such as a stream or gully, that forces the enemy to stop or slow down.

After the enemy has been ambushed and destroyed, the unit quickly withdraws over a prearranged route to the objective rally point. Speed is very important, since the noise of the ambush could alert other nearby enemy units.

13006. SECURITY PATROLS

General

Security patrols are assigned missions that may or may not require them to engage the enemy. They are used in proximity to defensive positions, on the flanks of advancing units or in rear areas. Purposes of security patrols are to detect infiltration by the enemy, destroy infiltrators, and protect against surprise and ambush.

In any situation where there is a threat of attack, such as a rear area threatened by guerrillas or a facility that is under threat of a terrorist attack, all Marines, not just the infantry, must know how to conduct a security patrol.

In just the offensive operations, infantry units provide security patrols to screen their flanks, areas, and routes. Whereas, in defensive operations, security patrols are used to prevent the enemy from infiltrating an area, detect and destroy infiltrators, and prevent surprise attacks. In rear areas, particularly when there is guerrilla or terrorist threat, the requirement to conduct security patrols increases for all Marine air-ground task force (MAGTF) units ashore, particularly aviation and combat service support units.

Task Organization and Equipment

Generally, a Marine rifle squad or similar sized organization is considered ideal for security patrols. Communications are important to higher headquarters so that they receive information from the patrol; and communications are important to the patrol to request fire support, etc. The radio the patrol carries must have the range necessary for higher headquarters to be able to receive transmissions from anywhere along the patrol route, and the patrol must have a secondary means of reporting (i.e., flare signals upon contact).

Patrol Procedures

All of the procedures presented in previous sections are to be used in security patrols.

Patrol Planning

Security patrol planning includes—

- Rehearsing prior to departing friendly lines.
- Maintaining communications.
- Support by organic weapons.
- Reinforced if necessary.
- Using varied routes and never establishing a routine pattern.
- Staying within proximity of friendly units.

Patrol Techniques

Within rear areas, an irregular pattern of patrol is established and changed daily. Outside of friendly lines it would be prudent to establish a definite preplanned route for the patrol, of which all adjacent units know the route. The parent unit commander establishes frequent checkpoints for control. If checkpoints are designated, the patrol leader treats them as individual objectives to be searched and cleared.

The patrol has a definite plan as to what to do if contact with enemy is made, how to break contact, how to defend itself, and how to call for supporting fires. It is imperative that patrol members know what to do if they become split or separated; i.e., location of rally points and how to be recovered.

13007. URBAN PATROLS

General

As national strategy continues to focus on regional vice global conflicts, the Marine Corps will continue to conduct urban operations in various operational environments. Cities and towns are often the center of economic and political power and are therefore extremely vulnerable to urban insurgent activities and violence. The lessons learned from recent operations in Somalia, as well as experiences gained by British forces in Northern Ireland, provide the foundations of urban patrolling.

Enemy Forces

Enemy forces in urbanized areas range from organized military forces to low intensity engagements with insurgents, such as terrorists or local gangs. For Marine Corps doctrine and tactics, techniques, and procedures (TTP) for dealing with the higher intensity threat in urbanized areas, refer to MCWP 3-35.3, *Military Operations on Urbanized Terrain.*

Common Threat Tactics

Cities provide cover and concealment for both friendly forces and enemy forces. However, enemy forces generally find active support only in certain areas of the town or city. The urban-based insurgent or terrorist usually lives in a friendly community or in one where the people are too frightened to withhold support or inform anyone about the situation. The insurgent or terrorist normally maintains close contact with leaders and others friendly to the cause. The enemy will often have efficient communication and intelligence systems, sometimes involving women and children to provide cover for its activities.

The urban insurgent normally cannot, like his rural counterpart, establish bases and recruit large military units. He is generally an individual or a member of a relatively small group. He relies on the cover afforded by the city's populace and terrorizing them to coerce loyalty or support.

Urbanized areas tend to give the insurgent and/or terrorist many opportunities to initiate action and gain advantage. The normal presence of large numbers of people in cities provides the foe an opportunity to mass crowds quickly and manipulate demonstrations easily. The presence of women and children during mass demonstrations may restrict the courses of action available to friendly forces. Major incidents stemming from overreaction or excessive use of force by friendly forces may provide the insurgent with propaganda material. Publicity is easily gained in an urbanized area because major incidents can't be completely concealed from the local population. Insurgent successes can be exploited to discredit the ability of host nation police, friendly forces, and the civil government, and gain recognition for the insurgents' cause.

The urban insurgent or terrorist can usually be expected to operate more boldly than his rural counterpart. This is reflected in the enemy's tactics. A single sniper or bomber may be the norm in the urbanized area, whereas the rural threat is generally the more conventional ambush. In urbanized areas, explosive devices can be easily emplaced and used effectively against large groups or select individuals. Enemy forces may be expected to employ the following tactics in urbanized areas:

- Using local communications, such as radio and newspapers, for propaganda purposes.
- Disrupting industry and public services through strikes and sabotage.
- Generating widespread disturbances designed to stretch the resources of the security force.
- Creating incidents or massing crowds in order to lure the patrol or reaction force into a trap.
- Provoking security forces in the hope that they may react improperly, therefore discrediting the security force by means of propaganda.
- Sniping at roadblocks, outposts, sentries, and patrols.
- Attacking friendly bases with rockets and mortars.
- Planting explosive devices, either against specific targets or indiscriminately, to cause confusion and destruction, and lower public morale and confidence.
- Using ambush patrols.
- Firing on friendly helicopters.

Principles of Urban Patrolling

Patrolling in an urban environment often presents conditions considerably different and often more complex than those encountered in rural and less inhabited areas. While the principles of patrolling are still relevant in an urban situation, the nature of urban patrolling has led to the development of six specific urban patrolling principles. They are—

1. **Depth.** The restrictive, canalizing nature of urbanized terrain usually limits a patrol's ability to disperse laterally. To prevent the patrol from bunching up, patrols normally maintain dispersion along the length of a patrol formation.

2. **Mutual Support.** The positioning of units in-depth within the patrol enables one unit to cover another unit's movement and facilitates immediate action during various situations. Aircraft, vehicles, and snipers also provide good mutual support.

3. **Deception and Pattern Avoidance.** Deception and pattern avoidance are normally a planning consideration of the headquarters directing the entire patrolling plan. By varying patrol routes, durations, and departure times, hostile actions commonly used against urban patrols—such as enemy ambushes, and roadblocks—are more difficult to plan and may be preempted.

4. **Intra-patrol Communication.** Elements of an urban patrol must have the means to communicate with each other. Ideally, each element will possess a radio enabling it to remain in continuous communication that facilitates rapid response and reporting to higher headquarters, reaction force coordination, and coordination of actions with other patrols or fire support agencies.

5. **Establishment of a Reaction Force.** The nature of urbanized terrain (its compartmentalization) makes urban patrols more vulnerable to a wider range of hostile actions. The requirement for immediate, coordinated reinforcement of a patrol is best satisfied by employment of an established reaction force. The reaction force requires superior mobility (relative to the enemy's) and fire support to be effective for this task.

6. **Three-Dimensional Threat.** Patrolling in an urban environment requires constant attention to its three-dimensional aspect; hostile actions can originate from rooftops, streets, subsurface levels or combinations of all levels at once.

Classification of Urban Patrols

Mission

The vast majority of urban patrols are overt in nature, with their presence readily apparent to the local populace. Most urban patrols are combat vice reconnaissance patrols. The vast majority of urban patrols are security patrols. However, units may be assigned secondary tasks of reconnoitering specific or general areas along the patrol routes.

Raids normally involve a swift penetration of an objective to secure information, confuse the enemy or destroy installations. Raids include a planned withdrawal upon completion of the assigned mission.

Movement

Means of movement are as follows:

- Dismounted: movement on foot.
- Vehicular: movement by motorized, mechanized or armored vehicle.
- Helicopterborne: movement by helicopter, however, helicopterborne patrols will usually involve dismounted or vehicular movement after patrol insertion.
- Combination: movement using a combination of methods.

Task-Organization

Units task-organize to meet the specific requirements of the mission and situation. The Marine rifle squad is ideally suited for urban patrolling and can easily integrate attached specialists required by specific missions. Specialists who may accompany urban patrols include—

- Interrogator-translator team (ITT) and counterintelligence team (CIT) Marines.
- Explosive ordnance disposal (EOD) personnel.
- Members of host nation or allied military forces.
- Interpreters.
- Local community leaders.
- Local law enforcement officers.
- Public affairs personnel escorting media representatives.

Dismounted Patrol Organization

Patrols should maintain unit integrity (fire team and/or squad) in organizing elements for an urban patrol. Similar to reconnaissance patrols, the vulnerability of urban patrols necessitates that all elements must provide for their own security in addition to the combat aspect of the mission. Combined assault and security teams are an effective method to organize for all-around security.

Urban Patrol Base Operations

Urban patrols may operate from an established patrol base that may be located within the unit's assigned area of operations or an area designated for a patrol. The patrol base should be located in a building used exclusively for this purpose. Patrol bases may also be located within a larger site that houses other agencies, such as a higher headquarters' command echelon.

If located within a larger site, the patrol base will be included in the overall facility security plan. If isolated, the patrol base must consider the following security factors:

- External security:
  - Barrier plan.
  - Sentry posts.
  - Local security patrols.
- Internal security:
  - Covered positions for all Marines.
  - Contingency plan for hostile actions against patrol base.

The Reaction Force

The high probability of an urban patrol becoming involved in a hostile or volatile incident requires the establishment of a dedicated reaction force for rapid reinforcement, support or extraction of the patrol. Ideally, the reaction force is—

- Large enough and task-organized in a manner that it can meet and quickly defeat the expected threat. Reaction forces are normally tiered with a lead element (normally one-third of the unit's size) and a main body force (the other remaining two-thirds of the force). Reaction force response times routinely are determined in advance by higher headquarters.
- Ready to respond immediately.
- Motorized or mechanized and supported by close air support and other fire support.
- Familiar with the area of operations.
- Briefed on the patrol's plans and monitors the status of patrols in progress.
- Task-organized to be multimission capable.
- Able to communicate with the higher headquarters, fire support assets, patrol base, and the patrol.
- Controlled by higher headquarters, once employed.

Patrol Preparation

Planning

Higher headquarters will—

- Designate the area for patrol.
- Provide intelligence briefs and updates.
- Ensure liaison with allied forces and the civilian populace.
- Provide special equipment and personnel required for the mission (scout snipers, public affairs officer, interpreters, etc.)
- Provide urban maps, photos, terrain models as required.
- Consider deception and pattern avoidance when issuing mission.
- Prescribe rules of engagement (ROE).

Intelligence Brief

An intelligence brief is conducted by the S-2 officer or representative prior to a patrol conducting its mission. The brief adresses the situation relevant to the specific patrol (e.g., routes, areas, updated enemy situations).

Coordination

Higher headquarters will effect liaison with adjacent and allied forces, as well as civil authorities and other agencies, having a possible effect on the patrol. The patrolling unit generally follows the same procedures as those used during patrol planning and execution in a jungle or forest environment.

The Urban Patrol Order or Warning Order

The Urban Patrol or Warning Orders use the same format and considerations as noted in this publication for patrolling. They rely heavily on a detailed terrain model, photographs, and subterranean construction to ensure complete understanding of the plan.

Rehearsals

The limited size of the patrol base usually precludes the need for full-scale rehearsals. Immediate action drills, such as crossing danger areas, are rehearsed in as much detail as possible, despite the limited available space.

Inspections

Initial and final inspections are conducted in the same manner as other patrols. Attached personnel must be fully integrated into the patrol and familiar with the plan and unit standing operating procedures (SOPs).

Conducting an Urban Patrol

Movement

Individual and unit movement considerations are generally the same as those for other patrols. However, urban environments require consideration of additional factors. Because of these factors, an urban patrol leader should—

- Ensure that each movement within a patrol takes place under the observation or cover of another individual or element of the patrol.
- Know where cover can be taken in the event of a hostile incident or action.
- Be prepared for contact with civilians, especially children, during the patrol and be aware that they may intentionally attempt to distract patrol members.
- Expect the presence of vehicles (both moving and stationary) along the patrol route.
- Expect members of the patrol to be approached by dogs and what action to take if threatened.

Patrol Formations

Squad-sized Patrols. The need for immediate fire power outweighs the dangers of becoming canalized. In contrast to other types of patrols, the headquarters element of an urban patrol will normally locate at the lead of the patrol column. This allows the patrol leader greater flexibility through control of two combined assault and security (A&S) teams. The leaders of these elements tactically stagger their members on each side of the road (see fig. 13-1 on page 13-12).

A&S teams follow in trace of the headquarters element and maintain unit integrity on separate sides of the street. One unit will remain slightly to the rear to create a staggered interval between Marines on either side of the street. This allows A&S teams to take lateral routes in support of headquarters element without having to cross a street to do so.

Platoon-sized Patrols. Squads will generally travel abreast of each other, moving along parallel routes. The interval between squad-sized units and/or teams is situation-dependent, but is usually between 100 and 150 meters (roughly two city blocks; this often prevents visual contact between the units). The intent is to create less of a target to an aggressor, yet still allow the patrol to quickly react to an incident. Individuals within units or teams will move in a staggered column as in a squad-sized patrol. See figure 13-2, which is located on page 13-13).

Night

Night patrols will generally be at least squad-sized and will generally use the same formation as that for day patrols. At night, it may be necessary to close distances between individuals or elements to maintain control. Consideration should be given to the use of night vision devices and thermal weapons sights.

Navigation, Control, and Security Measures

The patrol leader is ultimately responsible for the navigation. The headquarters element normally functions as the base unit during movement. The designated navigator is normally assigned from within the headquarters element. City maps are often inaccurate or outdated; however, when used with aerial photographs and other navigational aids, they can be effective for urban navigation.

Checkpoints and phase lines should be related to major streets (alleys, buildings, bridges) for easy identification. Arrival at checkpoints and crossing of phase

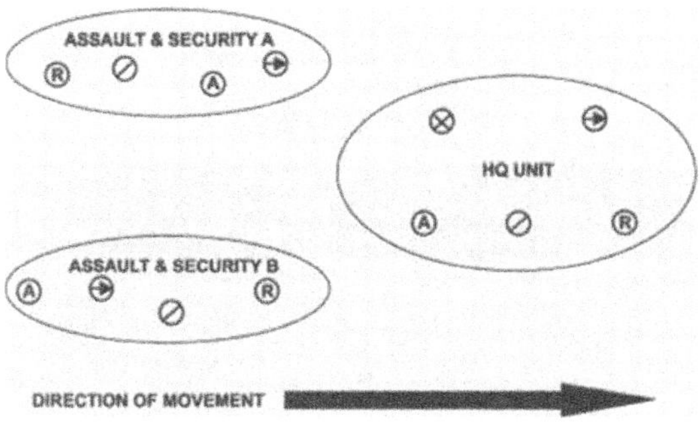

Figure 13-1. Squad-sized Dismounted Urban Patrol Formation.

lines should be relayed to higher headquarters using established brevity or codewords. A detailed patrol overlay is issued to both the reaction force and higher headquarters to keep them advised of routes and control measures used. A terrain model should be maintained at higher headquarters to aid in supervision and control of the patrol.

The use of camouflage should be limited to avoid frightening and confusing the local populace. The often overt nature of urban patrols may negate the need for camouflage. Patrols are sometimes deployed to show force presence and usually move on the urban street in plain view.

During daylight, patrols will routinely vary their rate of movement ranging from short halts to brief periods of double-timing. The British term for this urban patrolling technique is *hard-targeting*, meaning it makes the patrol harder for an enemy to target. Altering the rate of movement is intended to frustrate the enemy's ability to coordinate an attack or ambush against a targeted patrol.

Patrols should use short security halts, with Marines taking up mutually supporting firing positions. Marines must always work in pairs, ensuring mutual support. The last Marine in the element will provide rear security, but stays in his buddy's sight.

Individual Tasks

Individuals may be assigned collateral tasks performed throughout the patrol that may increase the patrol's situational awareness. Individual tasks may include—

- Vehicle spotter: looks for suspicious or known insurgent vehicles.
- Personnel spotter: observes and attempts to recognize previously identified enemy in crowds.
- Talker: attempts to gain information from casual conversations with the local populace. (Talkers are usually subordinate leaders or Marines with foreign language skills.)
- Searcher: conducts physical searches of vehicles and personnel while other patrol members provide cover and security.
- Marksman: engages point targets when the tactical situation does not permit massed or high volume fire. Other patrol members provide security to cover the marksman's engagement.

Departure of Friendly Lines

Urban patrols must vary their departure times to prevent being ambushed while exiting the patrol base.

Individual elements will usually depart exits at staggered times and at different movement rates, especially where sentries cannot provide cover. (This technique is used when the environment adjacent to the patrol base is dangerous.)

Once an element has exited friendly lines, a short halt is conducted in a predetermined, covered initial rally point 50 to 100 meters from the base. This ensures all elements are in position before the patrol continues.

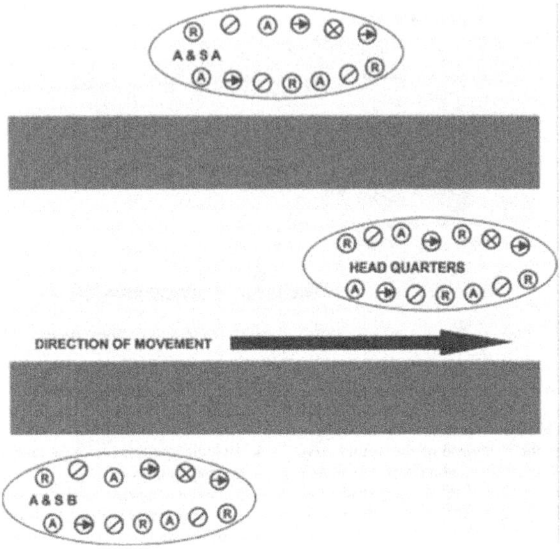

Figure 13-2. Platoon-sized Dismounted Urban Patrol Formations.

Exit points, routes from the base, departure techniques, and locations of IRPs should be varied constantly to avoid setting a pattern. This pattern avoidance may also include using vehicles to insert patrols away from the urban patrol base and employing empty vehicles as part of a deception plan.

Danger Areas

Urban patrols may encounter hundreds of danger areas during a single patrol. The three-dimensional threat requires keen situation awareness by every patrol member. Many danger areas can be dealt with simply by avoidance, while others require an adjustment of patrol formation, movement rate, etc. In the urban environment, places to be treated as danger areas are points that pose a major threat to the patrol, such as local political and religious headquarters, weapons containment areas, roads and routes that canalize movement and direct fire, and any area with a history of repeated contact.

Near and far side rally points are designated and briefed during the issuance of the patrol order. Squad-sized formations may use the A&S teams to provide flank security for the headquarters element and for each other. The headquarters element identifies the danger area and takes up a position on the near side of intersection. Individuals provide all-around security (see fig. 13-3 on page 13-14).

Two Marines (one from each A&S team) are designated to move through the headquarters element and establish respective firing positions on the near side of the danger area covering the patrol's near side flanks. They are followed by a second pair (again, one Marine from each A&S team) that moves across to the far side of the danger area and establishes respective firing positions covering the patrol's far side flanks (see fig. 13-4 on page 13-14).

Once near and far side flank security is established, the headquarters element moves across to the far side of danger area (see fig. 13-5 on page 13-15).

The remaining A&S team members then cross the danger area and join the headquarters element on the far side (see fig. 13-6 on page 13-15).

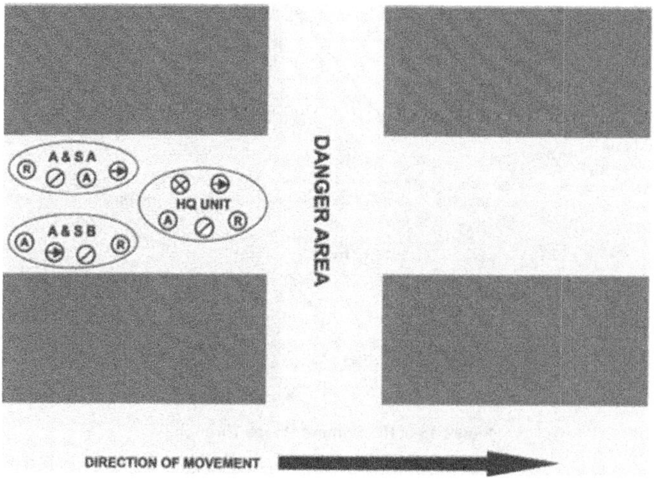

Figure 13-3. Approaching Danger Area.

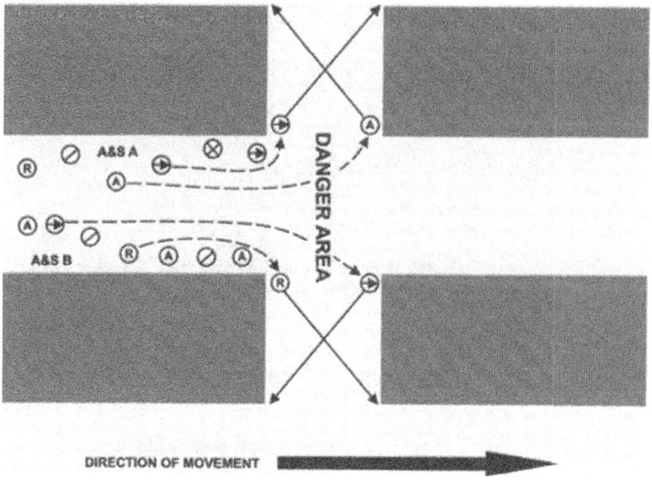

Figure 13-4. Securing Flanks.

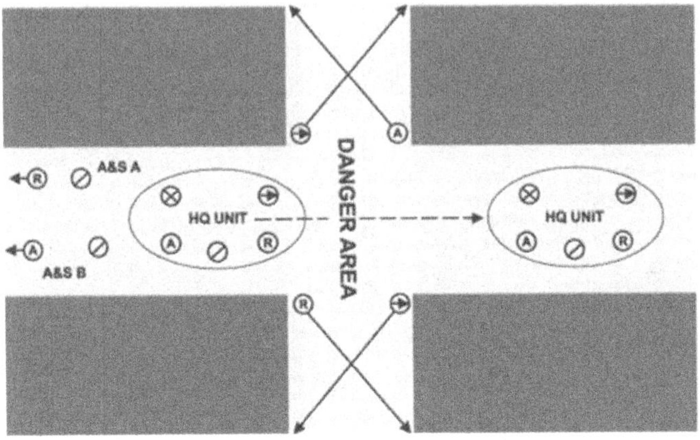

Figure 13-5. HQ Element Moves Through.

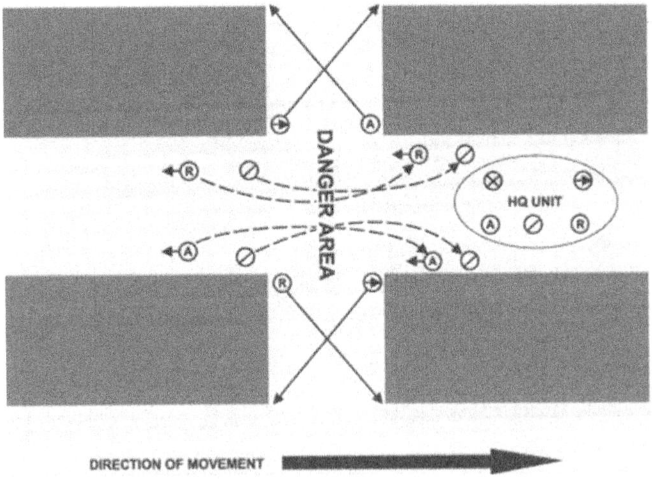

Figure 13-6. A&S Elements Move Through.

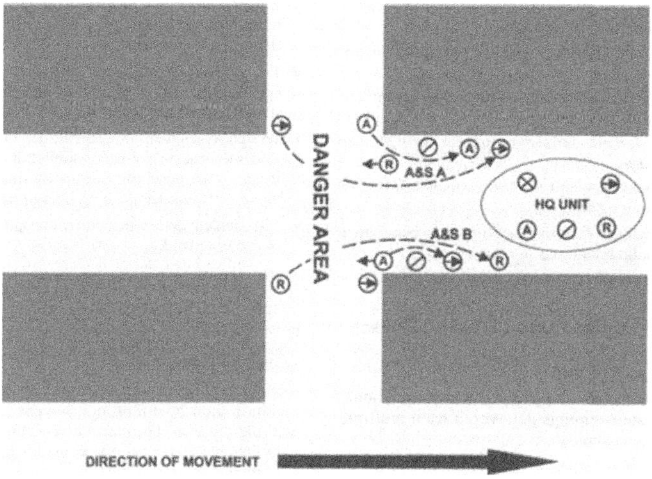

Figure 13-7. Patrol Resumes Movement Along Intended Route.

Once the trail A&S team members cross the danger area, they take up rear security and cover the movements of the near and far side flank security teams as they return to their positions in the patrol formation. The near side security team should collapse back first, followed by their far side counterparts. The patrol then resumes its advance away from the danger area (see fig. 13-7).

Interaction with Local Populace

Urban patrols must interact with the local populace. Patrols are at first a novelty to the civilians but can quickly become an unwelcome intrusion. The movement of the patrol must be fast enough to prevent the enemy from massing their fires upon it, but deliberate enough to ensure adequate security and mutual support. Patrol members must realize that they are usually the only Marines the local populace will encounter and that an inappropriate gesture, comment or act could lead to the deterioration of rapport between U.S. forces and the general population. Marines must remember that the vast majority of the individuals with whom they come in contact will be noncombatants attempting to survive in trying political, economic, and social situations.

Hostile incidents often seem to occur spontaneously, but there are usually indications that can alert Marines to imminent danger. The most obvious are the sudden alteration of normal routines, patterns, and attitudes of the local populace or other unusual activity. Some examples include—

- Observers on rooftops, in windows, etc., who are obviously tracking the patrol.
- The unusual absence of pedestrian traffic and people on porches.
- Stores, markets or street vendors closed suddenly or without explanation.
- Changes in civilian attitude toward patrol members.
- Unknown individuals or vehicles in the patrol area.
- Unfamiliar vehicles parked in the patrol area (possible car bomb).
- Roadblocks.
- Children throwing rocks at patrols to possibly draw the patrol's attention away from a more serious danger, such as a deliberate ambush.
- Vehicles riding unusually low due to overloading (possibly ferrying people, weapons, explosives).
- Agitators trying to provoke an incident with patrol members.

- Absence of the usual stray dogs (dogs are adept at sensing danger and avoiding it).
- Anti-American graffiti suddenly appearing in the patrol area.
- Pictures of enemy leaders and martyrs posted in the patrol area.
- Civilian workers failing to appear at U.S. or friendly bases.
- Normal deliveries and pick-ups conducted late or early without reason.
- Sudden change of civilian sentiment in newspaper articles, radio broadcasts or other media.
- Women and children leaving to live elsewhere.

Immediate Actions Upon Enemy Contact

Reaction to Sniping

Snipings are often executed from a single firing point, but coordinated snipings delievered from multiple points are not uncommon.

Patrol element or team leaders should constantly try to identify likely firing points and anticipate their own reactions to a possible shooting. Normally, snipers in an urban environment have a detailed withdrawal plan. Once a patrol comes into contact with a sniper, the patrol leader must immediately assess the situation and maneuver his patrol accordingly. The patrol's mission, location, size, ROE, and location of the threat often determine whether the patrol will attempt to neutralize the targeted sniper. If the patrol leader decides to kill or capture the sniper, he uses planned and rehearsed immediate actions to maneuver and counter the sniper's assault. The goal is to kill the sniper or cut off his escape and capture him. There are three immediate reactions to neutralizing a sniper: initial contact, immediate follow-up, and subsequent follow-up.

Initial Contact. The initial contact is made when the sniper fires the first shot. The patrol must react immediately and positively to get behind the firing position in order to kill or capture the gunman. The period of contact ends when the gunman is killed or captured, or the patrol element or team leader on the scene ends it. The following technique is the same for both squad- and platoon-sized patrols:

- The element or team in contact attempts to identify the firing position and maneuvers designated marksman into position to return well-aimed and controlled fire. Other members of the patrol take up positions to cover the marksman's engagement.
- The patrol element or team leader in contact sends initial contact report to the patrol leader, who notifies higher headquarters.
- The element or team leader in contact determines appropriate cut-off positions and relays them to flanking elements or teams.
- The patrol element or team leader in contact continues to observe the firing point, but does not enter it due to the possibility of booby traps. Flank elements or teams set up along likely escape routes.
- The incident ends when either the sniper ceases fire or is neutralized.

Immediate Follow-Up. Regardless of the fate of the gunman, isolation of the firing point is necessary to prevent reinforcement and preserve forensic evidence (scent, spent casings, etc.). If not under fire, members of the patrol element or team cordon off the area surrounding the firing point. Flanking elements or teams maintain their positions and prevent civilians from entering the area. The patrol leader moves to link-up with the element or team in contact (if not his own), and makes an estimate of the situation. The patrol leader sends a SPOTREP to higher headquarters.

Subsequent Follow-Up. The aim of the subsequent follow-up is to use follow-on forces to clear the building of remaining resistance or to obtain evidence that can be used to capture the gunman. The patrol leader establishes a position where he can brief arriving units (reaction force commander, S-2 representative, EOD personnel, etc.). Once the arriving units have been briefed, recommendations are made to higher headquarters via radio. No one is allowed into the cordon without the patrol leader's approval.

Reaction to Becoming Decisively Engaged

If a patrol becomes decisively engaged from numerous firing positions, the following immediate action should be taken:

- All patrol members move to available cover and return accurate fire on identified firing points.
- The patrol leader assesses the situation and makes a decision to either request the reaction force or break contact.
- If the reaction force is requested, the patrol will maintain its position until the reaction force arrives. The patrol should use fire and maneuver to gain better tactical positioning and support the arrival of the reaction force. When the reaction force arrives,

its commander may decide to either clear occupied buildings or cover the patrol during its extract.

Reaction to Bomb Threat or Discovery

The use of command-detonated explosive devices is a common ambush tactic employed by a terrorist or insurgent in an urban environment. The appropriate response to a reported threat or an actual discovery is generally involves four steps (known as the four Cs)—

1. The patrol leader **CONFIRMS** the presence of the suspicious item.

2. Without touching or moving anything suspicious, patrol elements or teams **CLEAR** the immediate danger area to a minimum of 100 meters. The area is cleared from the suspected device outward, inform civilians as to the reason for evacuation.

3. A&S teams establish a **CORDON** to secure the cleared area. Avenues of approach are cordoned off to keep people out and to protect EOD or engineer personnel clearing the device. The assistant patrol leader acts as the cordon commander and informs the patrol leader when the cordon is secure. An effective cordon technique is to tape off the area with engineer tape, creating both a physical and psychological boundary.

4. **CONTROL** of the area is maintained throughout the bomb clearing operation by the patrol leader. The patrol leader sends a report to a superior concerning details of the device (if known) and the area affected. The patrol leader coordinates with arriving personnel (EOD, engineers, etc.). The patrol leader maintains communication with the assistant patrol leader and keeps the Marines informed of the progress of the clearing operation.

Reaction to a Bomb Detonation

Bombs may be used by an insurgent as a means of initiating an ambush on mounted or dismounted patrols, in which case the actions for decisive engagement apply. Immediate action in response to an isolated explosion is similar to that used in reaction to a sniping and breaks down into the same three phases:

Initial Contact. The patrol leader attempts to identify the likely initiation point and sends an initial contact report to higher headquarters. If the bomb was command-detonated, the patrol leader sends his A&S teams deep to cut off the bombers' escape routes. Any casualties are moved a minimum of 100 meters from explosion and out of the line of sight to it.

Immediate Follow-Up. The A&S teams may need to maneuver to positions behind the likely initiation point to cut off escape. Once in position, personnel checks are conducted and any suspects are detained. The patrol leader coordinates requests for required support (MEDEVAC, reaction force, etc.)

Subsequent Follow-Up. Due to the possibility of secondary detonations, the four Cs (confirm, clear, cordon, and control) can be conducted as in reaction to a bomb discovery or bomb threat.

Civil Disturbances

Urban patrols must prepare to react to spontaneous aggression by the local populace. In many cases, civil disturbances are organized by the enemy to draw dismounted patrols into a targeted area, or to distract them from enemy activity occurring elsewhere. Civil disturbances are generally divided into two categories: minor aggressive actions, and full-scale rioting.

Minor aggressive actions are activities characterized by rock-throwing or use of devices such as Molotov cocktails and may either be directed at the patrol or take place between different ethnic factions of the population. Minor aggressive actions are normally spontaneous in nature and may have minimal or limited objectives for the insurgents.

Full-scale rioting events are usually in response to another major event or incident that may enflame the populace. Full-scale riots are well-planned and orchestrated, with clear objectives or targets in mind. At times, patrols will need to attempt to maintain control of a civil disturbance situation; however, dismounted and mobile small unit patrols should generally avoid potential flashpoints. Procedures to handle civil disturbances are as follows:

- The patrol leader reports the incident to headquarters and attempts to diffuse the crowd by talking to crowd leaders.
- If the patrol leader determines the size of the disturbance is too large for the force to handle, the patrol should move away from the disturbance to a safer, more remote covered area and occupy positions to observe and report the situation to higher headquarters. To prevent the patrol from being pursued by the crowd, the patrol should move quickly and change direction, often at road junctions, to gain distance from it.

Patrol members should maintain dispersion to create a more difficult target. They should face the crowd at all times to see and avoid any projectiles thrown. Individual self-discipline must be maintained throughout the disturbance. Marines charging into the crowd or throwing objects back at the crowd will only worsen the situation. If pursued or trapped, the patrol leader may consider using riot control agents (combat support, pepper gas, etc.) to disperse or slow a crowd's movement. The use of riot control agents must be authorized under the established ROE. If the patrol leader believes the situation is deteriorating beyond the patrol's ability to control it, the patrol leader should request the reaction force, which may be better equipped to handle a large riot or mob.

Break Contact

As with patrols in rural areas, the patrol leader may be forced to break contact as a result of decisive engagement with the enemy. On the basis of his estimate of the situation, the patrol leader will normally break contact in one of the following ways:

- As a patrol, with elements providing cover for movement as defined by clock direction and distance.
- As individual units/teams taking separate routes out of the area, then linking up at a designated rally point a safe distance away from the engagement.

As in any contact with enemy forces, smoke may be employed to screen movement. Fire support agencies can be utilized to suppress targets; riot control agents can be employed to disrupt enemy movement.

Reentry of Friendly Lines

The reentry of a dismounted patrol into an urban patrol base is no different from that of a patrol conducted in a rural area. The same planning considerations and control methods apply.

Missions Related to Urban Patrolling

House Calls

House calls missions are usually part of a coordinated effort to collect information within the area of operations. They involve obtaining up-to-date information on particular houses and occupants. When possible, local police should accompany patrols to do the actual talking to the occupants. If this is not possible, a technique that may be used by an urban patrol is—

- A&S elements move to provide cover around the target house.
- The headquarters element provides security just outside the house.
- The patrol leader and one Marine for security enter the house, if invited. If not invited, they talk to the occupants on the doorstep. Patience and tact are required in requesting information. An interpreter should be present when language differences exist.

Vehicle Checkpoints

Urban terrorists or insurgents commonly use vehicles to transport personnel, weapons, explosives, and equipment. Civilian vehicles are often used for these purposes, creating the requirement to check as many vehicles as possible. While permanent, fortified checkpoints may be conducted along approaches into an urbanized area, dismounted patrols can be employed to establish hasty vehicle checkpoints to stop vehicles and to keep the enemy off-guard. The two common types of vehicle checkpoints patrols establish are hasty and deliberate.

Hasty checkpoints are deployed anywhere based upon the decision of the patrol leader. Patrols must not set patterns through the frequent use of the same sites.

Deliberate checkpoints are tasked by higher headquarters to achieve a specific purpose. Time and locations are carefully considered to avoid setting patterns.

The general layout for a squad-sized, two-way dismounted checkpoint is depicted in figure 13-8 on page 13-20. The technique is as follows:

- The A&S teams are positioned stealthily in blocking positions on both sides of the road.
- Both the patrol leader and assistant patrol leader act as "talkers" for each direction of traffic (with local police or interrogator or translator Marines acting as interpreters) while a two-man team from the headquarters element physically searches the vehicles.
- Obstacles or parked vehicles may be employed to create a staggered roadblock in center of the checkpoint to slow approaching vehicles.
- The checkpoint location should be sited so that approaching vehicles cannot see it until they have passed a security team, and they have no escape route then available.
- Signs announcing the checkpoint should be displayed a safe distance from the search area for safety to both drivers and Marines.
- Normally higher headquarters will issue criteria that determines which vehicles are searched, but random checks of cars should normally be made as

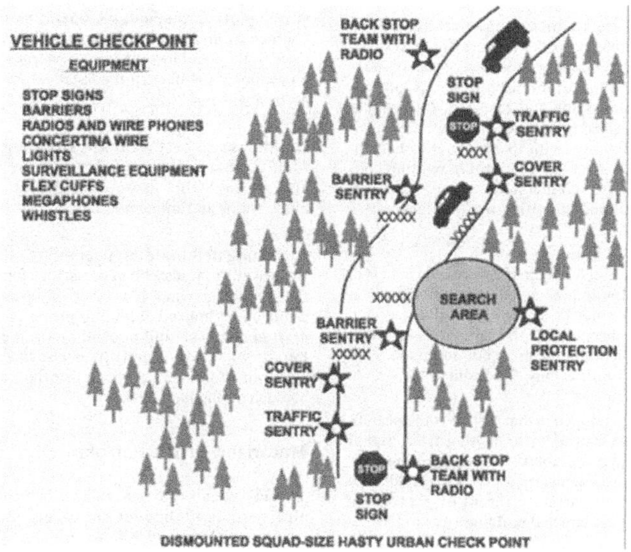

Figure 13-8. Dismounted Squad-sized Hasty Urban Vehicle Checkpoint.

well. While the vehicle is being searched the driver should accompany the searcher around the vehicle.

- Vehicle occupants should be made to exit the vehicle and then searched. Whenever possible, women should be searched by female police or female Marines.
- All patrol members should conduct themselves with courtesy and politeness. If nothing is found, an apology for the inconvenience is recommended.
- A hasty vehicle checkpoint should not be conducted any longer than 30 minutes for security reasons.
- The ROE should dictate whether or not action should be taken against vehicles that fail to stop at the checkpoint. Failure of a vehicle to stop does not automatically give authorization to fire.

Observation Posts

Urban observation posts are established to provide extended security, not only for patrol bases but also for patrols operating within the observation post's sector of observation. Observation posts can be established in conjunction with sniper operations and for providing overwatch for patrols operating within their sector of observation. Observation posts are nor-mally positioned on dominating terrain or in buildings outside the patrol base itself.

Insertion to the observation posts and conduct of observation may be either overt or clandestine in nature. Overt observation posts usually will be hardened positions to increase security. A patrol provides cover while the observation post is being inserted. A clandestine observation post relies on stealth of insertion and occupation for protection. It is normally positioned in abandoned buildings to cover sectors of observation that overt observation posts cannot. Because of their nature, clandestine observation posts are difficult to successfully establish and should not be manned for an extended period of time. Orders establishing observation posts (and patrols) must address the method of extraction as well as actions upon compromise/attack.

Cordon and Search

The cordon and search mission involves isolating a predesignated area by cordoning it off and systematically searching for enemy personnel, weapons, supplies, explosives or communications equipment. While large-scale cordon and search operations are planned and rehearsed in advance and normally entail extensive coordination with local law enforcement agencies, a squad-sized urban patrol may often conduct a cordon and search of a point target—searching one house or building identified by intelligence as a possible weapons cache.

The basic principle of a search of a populated area is to conduct it with limited inconvenience to the population. The populace may be inconvenienced to the point where they will discourage urban guerillas or insurgents from remaining in the area, but not to the point that they will assist the enemy as a result of the search.

Upon receiving intelligence that warrants the searching of a building or a specific tasking from higher headquarters, A&S elements of the patrol move to establish an inner cordon around the target building to seal it off, with the primary intent of preventing movement out of the targeted building.

On order, the designated reaction force deploys to establish an outer cordon, oriented outward some distance from the inner cordon and covering routes leading into the area in order to prevent outside interference/reinforcement. The reaction force maintains a reserve to reinforce either cordon or react to unfolding events (civil disturbance in response to the operation).

Once the cordons have been established, the patrol leader, with the assistance of local police or interpreters, informs the local populace that a building is about to be searched, that a house curfew is in effect (if permitted by higher headquarters), and that all occupants should remain indoors. Occupants of the target house are instructed to gather at a central location to stay out of the way of the search party.

The headquarters element, having linked up with any required assistance (explosive ordnance disposal (EOD), ITT, etc.) now acts as the search party and accompanies local police. A female searcher should be included in the party, if necessary.

Occupants are searched and screened first for possible enemy personnel. Apprehended persons are evacuated as soon as possible.

The head of the household should accompany the search party throughout the operation to be able to counter incriminating evidence and possible accusations of theft and looting against Marines. If possible, a prominent member of the local community should act as a witness.

Buildings are best searched from top to bottom. Ideally, the search is conducted with the assistance of combat engineers using mine detectors to locate hidden arms and ammunition.

If the targeted building is empty or the occupant refuses entry, it may be necessary to forcefully enter the premises to conduct a search of the dwelling. If an unoccupied house containing property is searched, arrangements should be made with the local community to secure it until its occupants return. Unnecessary force and damage to property should be avoided during the search.

Motorized Urban Patrols

The advantages of a motorized urban patrols is their ability to capitalize on the speed, mobility, and protection offered by various vehicles. They may be motorized, mechanized or armored vehicles or a combination. Generally, motorized urban patrols possess greater combat power than dismounted patrols and can cover larger areas faster than dismounted patrols.

The disadvantages of motorized urban patrols is that they are restricted to roads and are vulnerable to ambush by the enemy. They are also restricted in their ability to interact with the local populace.

Motorized patrols are generally organized in the same manner as dismounted patrols (see fig. 13-9). Unit integrity is maintained when assigning personnel to specific vehicles.

The urban patrolling principles apply to motorized patrols in much the same manner as dismounted patrols. Mutual support and depth are achieved by maintaining constant observation between vehicles and coordinating support with any dismounted patrols in the area. All-around security is achieved through the use of constant observation as well as the vehicle's mobility and firepower. Positive communications between units or teams are maintained through vehicle radios. Patrol routes and speeds are varied to promote deception or pattern avoidance.

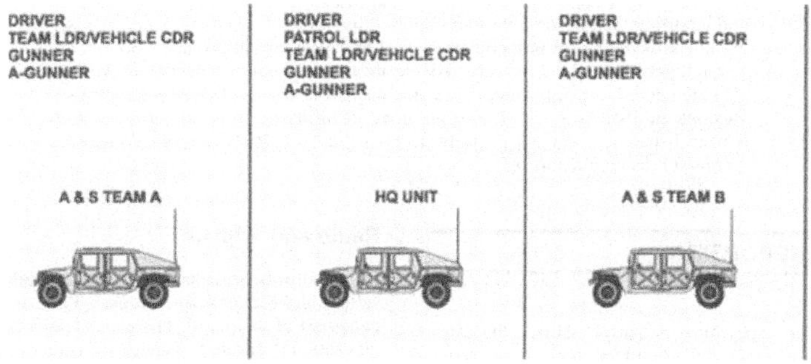

Figure 13-9. Organization of a Squad-sized Motorized Patrol.

The canalizing nature of streets and alleys force vehicular patrols to use a traveling overwatch movement technique to reduce vulnerability to ambushes. All vehicles travel at a moderate rate of speed with the lead vehicle stopping only to investigate potential danger areas. If vehicles must stop in danger areas, designated crew members' will dismount to provide security. The gunner will remain at the ready and in the turret while the driver remains in the driver's seat with the vehicle running.

Vehicles should move at a high rate of speed only when responding to an incident. At all other times, vehicle speed should be between 15 to 20 mph to allow for quick reaction and good observation. Distances between vehicles should be approximately 50 meters (one half to one city block) or such that visual contact and mutual support are ensured. Particular care is taken at major road junctions and other danger areas to ensure individual vehicles do not become isolated.

Vehicles with doors removed generally enhance observation and overall security, yet expose Marines to thrown objects, theft and concealment.

CHAPTER 14. INFORMATION AND REPORTS

It is necessary that patrol leaders and all patrol members be trained in observing and reporting their observations accurately. The leader of a patrol should have all members of the patrol immediately signal or report any information obtained. These reports should not be restricted to information about the enemy, but should also include information about the terrain, such as newly discovered roads, trails, swamps, and streams. The leader includes all information in the report to the officer dispatching the patrol. (Refer to MCRP 2-15.3, *Reconnaissance Reports Guide*, for detailed report formats.)

14001. REPORTING

The officer dispatching the patrol instructs the patrol leader on whether and when messages are to be sent back during the patrol and what communication means to use. Messages may be oral or written. They must be accurate, clear, and complete. Every message should answer the question what, where, and when. For a detailed discussion on reporting, see paragraph 6002.

Verbal Messages

A patrol leader sending a verbal message should make it simple, brief, and avoid using numbers and names. The messenger should accurately repeat the message back to the patrol leader before leaving.

Written Messages

In preparing written messages, the patrol leader must distinguish between fact and opinion. Information about the enemy should include: strength; armament and equipment; actions; location and direction of movement; unit destination, if known; time enemy was observed; and the patrol's location when the observation was made. Use of an overlay or sketch may often simplify the message.

Messengers

A messenger team is given exact instructions as to where to deliver the message and the route to take. Any information obtained along the route should be reported at the time the message is delivered. Messengers must be given all practical assistance. If in danger of capture, the messenger immediately destroys the message.

Radio and Signals

If the patrol is provided with a radio, a definite radio schedule for checking in must be arranged before departure of the patrol. The patrol leader takes every precaution to ensure that codes and copies of messages are not captured by the enemy. If a close reconnaissance of enemy lines is required, the radio should be left in a concealed location at a safe distance from the enemy. Once a report is sent by radio, the patrol should immediately leave the area to avoid the possibility of detection by enemy locating devices. Pyrotechnics (flares, colored smoke, grenades) and air-ground panels may also be used by patrols for reporting information by a prearranged signal.

SALUTE Report

Information must be reported as quickly, accurately, and as completely as possible. An established method to remember how and what to report about the enemy is to use the acronym SALUTE:

Size
Activity
Location
Unit
Time
Equipment

An example of such a report is: "Seven enemy soldiers, unit unknown, traveling SW, crossed road junction on BLACK RIDGE at 211300 August carrying one machine gun and one rocket launcher."

14002. CAPTURED ITEMS

Every patrol should make a practice of searching enemy casualties, prisoners, and installations first for booby traps, then for equipment, papers, maps, messages, orders, diaries, and codes. Search techniques should be practiced often in order to reduce the time exposed to potentially dangerous situations. Items found are collected by the patrol leader and turned in with the patrol report. The items found are marked as to time and place of capture. When possible, captured items should be linked to a specific prisoner who possessed the items or to the place where the items were found. When this is done, the enemy prisoner of war tag and item tag are marked accordingly. The patrol leader must impress upon the members of the patrol the importance of turning in all documents and equipment. Furthermore, the patrol leader must ensure all information gathered by the patrol is quickly disseminated.

14003. PRISONERS

A patrol normally does not capture prisoners unless required by the mission. If prisoners are taken, the "5S and T" rule applies. This memory aid stands for: search, segregate, silence, speed, safeguard, and tag.

Search. Prisoners are body-searched thoroughly for weapons and documents as soon as they have been captured. This search must include the helmet, body armor, and gas mask. These items are left with the prisoner for protection until the patrol is completed. Weapons, equipment, and documents are tagged and immediately sent to the patrol leader.

Segregate. Prisoners are segregated into isolated groups: officers, noncommissioned officers (NCOs), privates, deserters, and civilians. By segregating prisoners, it makes it more difficult for leaders to organize escapes and issue orders to subordinates.

Silence. Silence is essential. Do not allow prisoners to talk to each other.

Speed. Speed is required in getting prisoners to the commander who dispatched the patrol. Timely information secured from prisoners is essential.

Safeguard. Prisoners are safeguarded as they are moved. They are restrained, but not abused. If the patrol will soon reach friendly positions, prisoners are not given cigarettes, food or water until they have been questioned by interrogators. If the patrol will not return to a friendly position for a lengthy time, the prisoners are given food, water and medical aid.

Tag. Equipment and personal effects recovered from prisoners are tagged so that they may later be matched back to specific individuals.

14004. PATROL REPORT

Every patrol leader makes a report when the patrol returns. Unless otherwise directed, the report is made to the person ordering the patrol. If the situation permits, the report is written and supported by overlays and/or sketches. The patrol leader's report should be a complete account of everything of military importance observed or encountered by the patrol while on the assigned mission. It should include the following information:

- Size and composition of patrol.
- Tasks and purpose (mission).
- Time of departure.
- Time of return.
- Routes, out and back (show by sketch, azimuth, trace on map).
- Terrain (general description to include any man-made or natural obstacles and critical terrain features which, if occupied by either enemy or friendly forces, would allow them to control the surrounding area).
- Enemy (size, activity, location, unit, time, equipment).
- Any map corrections (show on map).
- Miscellaneous information not covered elsewhere in report.
- Results of enemy encounters.
- Condition of patrol, including disposition of any dead or wounded.
- Conclusion and recommendations.

14005. PATROL CRITIQUE

After the patrol has rested and eaten, the patrol leader should hold a critique. Constructive criticism is made. It is an excellent time to prepare for future patrols by going over lessons learned as a result of the patrol.

Appendix A. Patrol Warning Order

The warning order is issued as soon as practical with all available information included to assist patrol members in preparation.

1. Situation. Friendly and enemy situation information necessary for initial preparation.

2. Mission. Statement of what the patrol is to accomplish, and the purpose for accomplishing it. When, how, and where will be discussed in the patrol order.

3. Execution

a. Task Organization. General patrol organization and assignment of responsibilities if known; otherwise, promulgate in the patrol order.

b. Tasks. Alert subordinate leaders to patrol tasks (e.g., stream crossing, helicopter rappelling, demolitions) requiring preparation prior to departure.

c. Coordinating Instructions

(1) Time schedule and location for individual preparation, rest, briefings, in-spection, rehearsal, and departure.

(2) Time, place, uniform, and equipment for receiving the patrol order.

(3) Tasks for subordinate leaders to direct and supervise the initial preparation which may include drawing ammunition, rations, and special equipment; conducting immediate action drills or other necessary individual or unit training; meeting and briefing attachment personnel; reconnoitering the area for passage of lines; and coordinating with the necessary unit leaders.

(4) Preliminary guidance to specialists and key individuals regarding their roles and organization within the patrol.

4. Administration and Logistics

a. Individual uniform, equipment, weapons, and prescribed load of rations, water, and ammunition.

b. Crew-served weapons (if required) and guidance regarding distribution of weapons and ammunition during movement.

c. Special equipment requirements (wire cutters, demolitions, radios, flashlights, infrared equipment, mines, binoculars) and their distribution during movement.

d. Restricted or prohibited items.

5. Command and Signal

a. Designation of assistant patrol leader and his role in preparation.

b. Designation of navigators and radio operators (if required).

c. Brief outline of patrol leader's schedule for preparation, and where he can be reached.

Appendix B. Patrol Order

A patrol order follows a warning order. Any of the following subjects that have been addressed in the warning order may be omitted with the exception of the mission statement. A patrol order is more detailed than a 5-paragraph squad order in that a great deal of attention is given to individual duties.

The patrol order also provides orientation information, both enemy and friendly, that impacts the patrol and enemy forces. This information can include, but is not limited to, weather, terrain, visibility, NBC considerations, local population situation, terrain model and/or map orientation, and behavior.

1. Situation

a. Enemy Forces

(1) Composition, disposition, and strength are based on size, activity, location, unit, time, and equipment (SALUTE).

(2) Capabilities and limitations to defend, reinforce, attack, withdraw, and delay (DRAW-D).

(3) Enemy's most probable course of action.

b. Friendly Forces

(1) Mission of next higher unit (task and commander's intent).

(2) Adjacent unit missions (task and intent). Also identify left, front, right, and rear.

c. Attachments and Detachments. (date and time effective).

2. Mission. The mission is the task to be accomplished, and its purpose (who, what, where, when, and why). For patrols, specify if the mission or time has priority.

3. Execution

a. Commander's Intent and Concept of Operations

(1) Commander's intent.

(2) The concept of operations tells the where, how, and who and lays out the patrol leader's general scheme of maneuver and fire support plan. It outlines the following:

(a) Task organization of the patrol.

(b) Movement to the objective area, to include navigation method.

(c) Actions in the objective area.

(d) The return movement, to include navigation method.

(e) Use of supporting forces (including illumination, if required).

b. Subordinate Element Missions. Subordinate element missions (task and purpose) are assigned to elements, teams, and individuals, as required.

c. Coordinating Instructions. This paragraph contains instructions common to two or more elements, coordinating details, and control measures applicable to the patrol as a whole. At a minimum, it includes—

(1) Time of assembly in the assembly area.

(2) Time of inspections and rehearsals (if not already conducted).

(3) Time of departure and estimated time of return.

(4) Location of departure and reentry of friendly lines and the actions associated with departure and reentry.

(5) Details on the primary and alternate routes to and from the objective area.

(6) Details on formations and order of movement.

(7) Rally points and actions at rally points.

(8) Final preparation position and actions at this position.

(9) Objective rally point and actions at this point.

(10) Actions at danger areas.

(11) Actions in the event of enemy contact.

(12) Details on actions in the objective area not covered elsewhere.

(13) Estimated time of patrol debriefing upon return.

4. Administration and Logistics

a. Changes/additions to uniform, equipment, and prescribed loads from that given in the warning order.

b. Instructions for handling wounded and prisoners.

5. Command and Signal

a. Command Relationships. Identify key leaders and chain of command.

b. Signal. Challenge and password, arm and hand signals, special signals, and radio frequencies and call signs.

Appendix C. Patrol Evaluation Checklist

This appendix provides a comprehensive checklist of critical patrolling steps, techniques, and procedures to aid unit leaders to critique patrol performance during training. Unit leaders may use the list as they observe the performance of a patrol to provide the patrol leader with a detailed analysis of the performance.

Warning Order	Yes	No
Did the PL ensure all patrol members were present before issuing the warning order?		
Did the PL issue a brief statement on the enemy situation?		
Did the PL issue a brief statement on the friendly situation?		
Did the PL state the mission in a clear, positive manner?		
Did the PL list all members of the patrol including attachments?		
Was the chain of command for the patrol covered fully?		
Were all members of the patrol asigned positions and duties in particular squads and teams?		
Were all the necessary individual duties assigned?		
Did the PL follow established principles in organizing the patrol into squads and teams?		
Was each patrol member assigned a particular weapon to carry on patrol?		
Did the PL list all special equipment needed to accomplish the mission?		
Was required special equipment assigned to the proper element to carry?		
Did the PL select uniform and equiment common to all based on METT-T?		
Coordination with Adjacent Units		
Did the PL coordinate with other patrols operating to the right and left?		
Was the route out and back coordinated?		
Was the time of departure and return cocordinated?		
Were call signs and frequencies coordinated?		
Was a signal for the FPF coordinated so as not to approach friendly lines during this time?		
Coordination With Front Line Units		
Did the coordinator pass the size of the patrol?		
Was the time of departure and return coordinated?		
Did the coordinator give a general area of oeprations for the patrol?		
Did the coordinator ask for information on known or suspected enemy positions and/or obstacles?		
Did the coordinator ask about information on the latest enemy activity?		
Did the coordinator ask for detailed information on friendly fire support available and the unit's barrier plan?		
Was the location of the IRP established and coordinated?		
Did the coordinator ask the forward unit to monitor their patrol frequency?		
Was the current challenge and password confirmed?		
Did the coordinator request that all information coordinated be passed on to any relieving unti?		
Patrol Order		
General		
Did the PL check to ensure all patrol members were present before issuing the patrol order?		

	Yes	No
Did the PL issue the patrol order in a forceful, confident manner?		
Did the PL make maximum use of available visual aids in issuing the patrol order (i.e., terrain model, sand table, map board, chalkboard)?		
Did the PL issue the patrol order in correct sequence?		
Did the PL issue the entire patrol order without allowing interruptions by patrol members?		
Did the PL adequately answer all questions asked by patrol members		
Did paragraph 1a (Enemy Situation) include—		
A weather forecast for the period of operation?		
A description of the terrain over which the patrol was to operate?		
Identification or description of enemy units known to bein the area of operations?		
Known locations of enemy units?		
Recent activity of enemy units?		
Stength of enemy units in the operating area?		
Did paragraph 1b (Friendly Situation) include—		
Mission of the next higher unit?		
Locations, missions, and planned actions of units on right and left?		
Fire support available to support the patrol (as per the coordination)?		
Missions and routes of other patrols operating in the immediate area?		
Did paragraph 1c (Commander's Intent) include—		
All views and ides of what the commander wants to be accomplished?		
The final result desired by the commander?		
Did paragraph 1c (Attachments and Detachments) include—		
All attachments to patrol and effective time of attachment?		
All detachments from patrol and effective time of detachment?		
Did paragraph 2 (Mission) include, at a minimum—		
Who was to conduct the patrol?		
What the patrol was tasked to do (i.e., conduct a point recon patrol)?		
Where the action was to take place?		
Why the action was to take place (i.e., the purpose)?		
Did paragraph 3 a (Commander's Intent and Concept of Operations) include—		
A complete concept of operations?		
A detailed description of the mission of elements?		
A detailed description of the mission of teams?		
A detailed description of the duties of specific individuals (i.e., navigator, compass man, pace m an, corpsman, APL)?		
Time of departure and time of return?		

	Yes	No
Type of formations and order of movement to be used?		
Description of the route and alternate routes(s) to include azimuths and distances between checkpoints as a minimum?		
Techniques to be used in the departure from friendly areas?		
Techniques to be used for the reentry into friendly areas?		
Location of the IRP and a tentative ORP, to include grid coordinates and recognizable terrain features?		
Method for designating and passing rally points?		
Actions to be taken at rally points if their use became necessary?		
Actions to be taken in the event of enemy contact?		
Actions to be taken at danger areas?		
Complete, detailed description of actions to be taken at the objective?		
Times and locations for rehearsals and the order of priority for rehearsals?		
Times and lcoations for inspections and the methods of conduct (to include uniform and equipment to be worn and/or carried)?		
Where, when, and by whom the debriefing is to be conducted?		
Did paragraph 4 (Administration and Logistics) include—		
Rations to be carried or reference to the warning order if there is no change?		
Arms and ammunition to be carried or reference to the warning order if there is no change?		
Uniform and equipment to be worn and carried or reference to the warning order if there is no change?		
Method for handling wounded or dead?		
Method for handling prisoners?		
Did paragraph 5a (Signal) include—		
Complete description of all signals, code words, frequencies, and call signs to be used within the patrol?		
Were signals adequate?		
Radio call signs to be used?		
Primary and alternate frequencies to be used?		
Call signs and frequencies to be used with other units in the area of oeprations (if applicable)?		
Required reports to be transmitted to higher headquarters, along with occasions for reporting?		
Code words and brevity codes to be used between the patrol and higher headquarters?		
Current challenge and password to be used in friendly held areas?		
Current challenge and password to be used forward of FEBA?		
Did paragraph 5b (Command) include—		
Chain of command or reference to the warning order if there is not change?		
Location of the PL during all stages of operation?		

	Yes	No
Location of the APL during all stages of operation?		
Inspection Rehearsal		
General		
Did the P L adequately use the allotted inspection time?		
Was the inspection conducted in an orderly manner?		
During the inspection, did the PL check the following:		
Completeness and correctness of uniform?		
Completeness of all equipment necessary to accomplish the mission?		
Operational condition of equipment?		
Did the PL question members to ensure they knew—		
The mission of the patrol?		
The concept of operations?		
Their individual duties and responsibilities?		
Chain of command and succession?		
Duties and responsibilities of key personnel?		
Did the PL utilize the rehearsal area to conform as much as possible to area of operations?		
Were the following major actions rehearsed:		
Actions at the objective?		
Actions at danger areas?		
Actions on enemy contacts?		
Departure and reentry of friendly lines?		
Did the PL critique each action after it had been rehearsed?		
Were interpatrol communications and control measures checked at rehearsal?		
Did the PL maintain control of the patrol during the rehearsal?		
Movement		
Did the PL use proper formations for movement?		
Did the PL adequately control the point team?		
Did the PL check the compass men?		
Was the PL aware of his position at all times?		
Was light and noise discipline enforced?		
Was the rate of movement appropriate for the mission?		
Did the PL make full use of pace?		
Did the PL make full use of count?		
Was proper security maintained during movement?		
Did the PL select and pass proper rallying points?		
Did the PL ensure that all members knew location of rallying points?		

	Yes	No
Did the PL properly use arm and hand signals?		
Did the PL recognize and halt the patrol a safe distance from a danger area?		
If the PL conducted a map check, did he conduct it properly (light discipline, security, etc.)		
Did designated individuals reconnoiter the far side of a danger area properly?		
Were support teams emplaced properly?		
Did the PL use a proper formation in crossing the danger area?		
Were reports made on enemy contact and at checkpoints?		
Actions on Enemy Contact		
Did the PL halt patrol a safe distance from a tentative ORP?		
Did the PL issue an adequate frag order before departing to look for a tentative ORP?		
Did the PL take appropriate personnel to look for the ORP?		
Did the PL select a suitable ORP?		
Did the PL secure the site adequately?		
Did the patrol occupy the ORP as stated in the patrol order or subsequent frag order?		
Did the patrol occupy the ORP in an orderly manner?		
Did the PL maintain control of the patrol during occupation of the ORP?		
Did the PL issue a satisfactory contingency plan to the APL before departing on a leader's recon?		
Did the PL take appropriate personnel on a leader's recon?		
Did the PL maintain or ensure that the patrol leader had communications with the patrol?		
Did the patrol avoid being detected by the enemy during the leader's recon?		
If contact was made, did the PL take appropriate action?		
Was security maintained during the leader's recon?		
Did the PL ensure the objective was kept under surveillance?		
Were appropriate orders given to surveillance teams?		
If a leader recon proved a tentative ORP unsuitable, did the PL move the ORP?		
Was a satisfactory leader's recon conducted?		
Was the ORP move conducted in a satisfactory and orderly manner?		
On return to the ORP, did the PL issue appropriate frag order and allow enough time for dissemination?		
Actions at Objective		
Did the PL issue a frag order for action at the objective?		
Was security emplaced prior to the teams departing from the ORP?		
Did the PL employ the terrain at the objective to the best advantage?		
Was stealth maintained while moving into the objective?		

	Yes	No
Did the Pl exercise positive control of squads, teams, and individuals at the objective?		
Were actions at the objective in accordance with the details outlined in the patrol order?		
Did the PL make use of supporting arms at the objective?		
Was the action at the objective successful?		
Was withdrawal from the objective accomplished quickly and orderly?		
Did units withdraw according to the details reflected in the patrol order?		
Was reorganization at the ORP completed in an expeditious and orderly manner?		
Did the patrol withdraw from the ORP quickly and quietly?		

Appendix D. Ambush Formations

Formations discussed here are identified by names which correspond to the general pattern formed on the ground by deployment of the assault element.

Line Formation

The assault element is deployed generally parallel to the route of movement of the enemy. This positions the assault element parallel to the long axis of the killing zone and subjects the target to heavy flanking fire. The size of the force which can be trapped in the killing zone is limited by the area the assault element can effectively cover. The enemy is trapped in the killing zone by natural obstacles, mines, demolitions, and direct fires. (See fig. D-l.)

A disadvantage of the line formation is the chance that lateral dispersion of the target may be too great for effective coverage. The line formation is appropriate in close terrain that restricts enemy maneuver, and in open terrain where one flank is protected by natural obstacles or can be protected by mines and demolitions. Similar obstacles can be placed between the attack force and the killing zone to provide protection from possible enemy counter attack. When an ambush is deployed in this manner, access lanes are left so that the enemy can be assaulted. (See fig. D-2.)

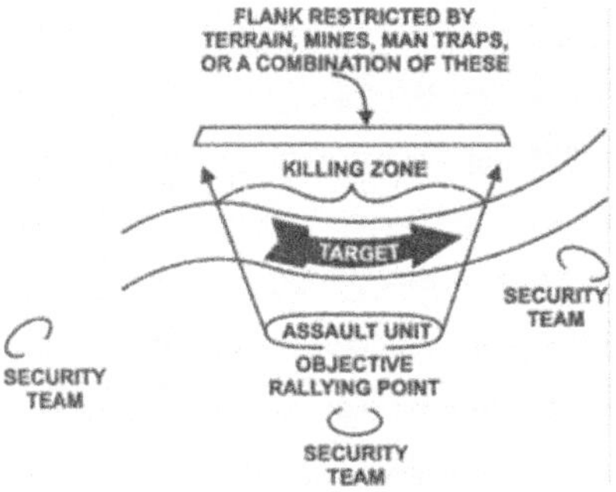

Figure D-1. Line Formation (Harassing or Destruction Ambush).

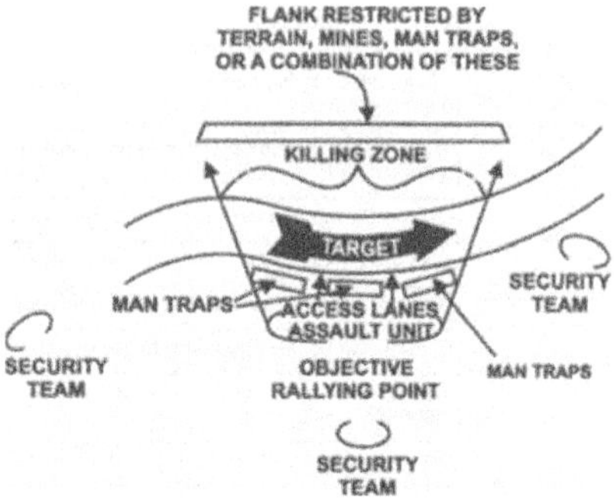

Figure D-2. Line Formation (Access Lanes for Assault of Target).

The main advantage of the line formation is its relative ease of control under all conditions of visibility.

L Formation

The "L" shaped formation is a variation of the line formation. This formation is very flexible because it can be established on a straight stretch of a trail or road (see fig. D-3) or at a sharp bend in a trail or a road (See fig. D-4 on page D-2).The long side of the assault element is parallel to the killing zone and delivers flanking fire. The short side of the attack force is at the

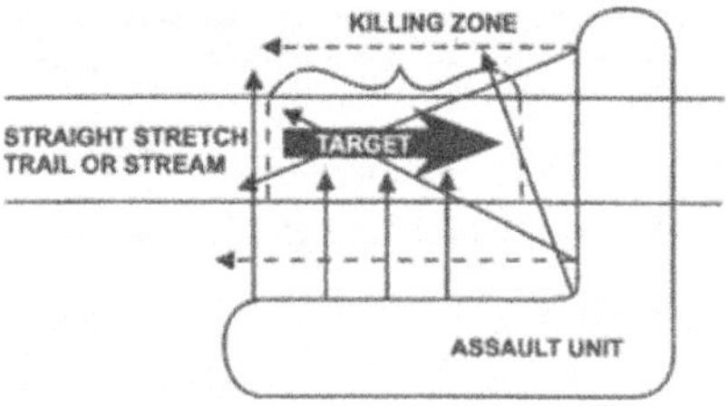

Figure D-3. "L" Formation (Destruction Ambush).

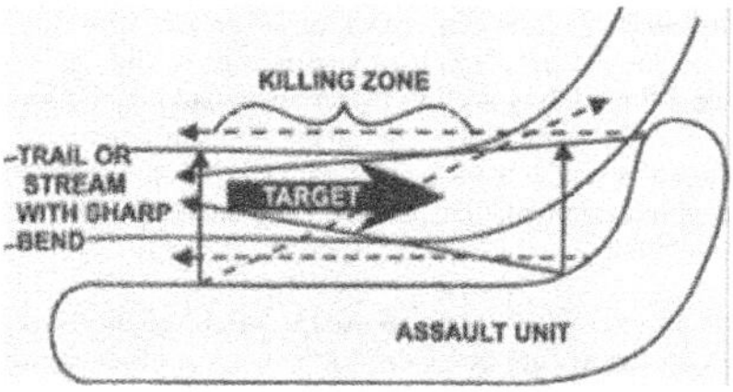

Figure D-4. "L" Formation (Bend of Trail or Stream).

end of, and at right angles to, the killing zone and delivers enfilading fire that interlocks with fire from the long side. When appropriate, fire from the short side can be shifted to parallel the long side if the enemy attempts to assault or escape in the opposite direction. In addition, the short side prevents escape and reinforcement. (See fig. D-5.)

"Z" Formation

The "Z" shaped formation is another variation of the line formation.

The assault element is deployed as in the "L" formation, but with an additional side so that the formation resembles a "Z". (See fig. D-6).

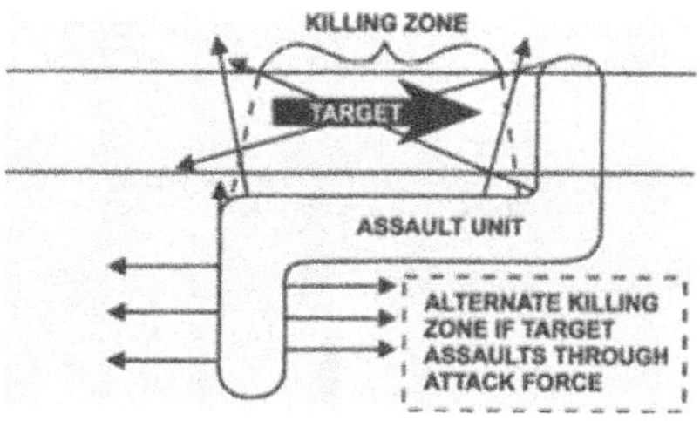

Figure D-6. "Z" Formation.

The additional wing may serve any of the following purposes:

- To engage an enemy force attempting to relieve or reinforce the enemy unit engaged in the kill zone.
- To seal the end of the killing zone.
- To restrict a flank.
- To prevent an envelopment.

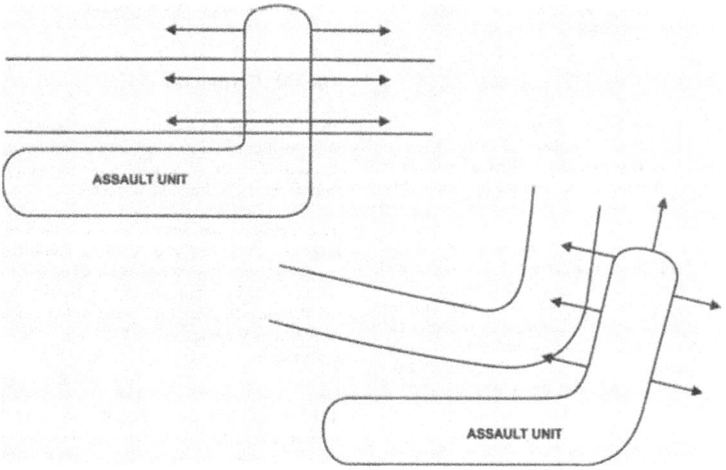

Figure D-5. "L" Formation.

"T" Formation

In the "T" shaped formation, the assault element is deployed across (perpendicular to) the enemy's route of movement so that its position forms the crossing of a "T" at the top. (See fig. D-7.)

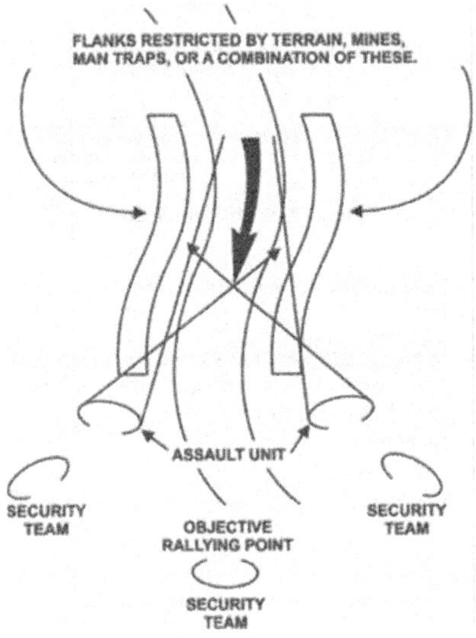

Figure D-7. "T" Formation.

This formation can be used day or night to establish an ambush to interdict movement through open areas that are hard to seal off.

A small force can use the "T" formation to harass, slow, and disorganize a larger force. When the lead elements of the enemy are engaged, they will normally attempt to maneuver right or left to close with the ambush. Mines and other obstacles placed to the flanks of the killing zones slow the enemy's movements and permit the ambush force to deliver heavy fire and withdraw without becoming decisively engaged.

The "T" formation can be used to interdict small groups attempting night movement across open areas. For example, the assault element is deployed along an avenue of approach with every second man facing the opposite direction. The attack of the enemy approaching from either direction requires only that every second man may shift to the opposite side of the formation. Each man fires only to his front and only when the enemy is at a very close range. Attack is by fire only and each man keeps the enemy under fire as long as it remains to his front.

If the enemy attempts to escape in either direction along the killing zone, each Marine takes the enemy under fire as the enemy comes into the Marine's sector of fire. The "T" formation is very effective at halting infiltration. But it has one chief disadvantage; there is a possibility that the ambush will engage a superior force at night while spread out. (See fig. D-8.)

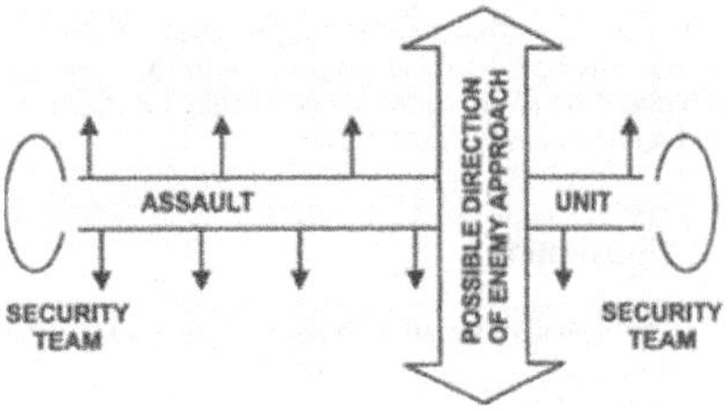

Figure D-8. "T" Formation (Target Approaching from Either Direction).

The "V" Formation

The "V" shaped formation is deployed along both sides of the enemy's route of movement so that it forms a "V"; care is taken to ensure that neither group (within the "V") fires into the other.

This formation subjects the enemy to both enfilading and interlocking fire. The "V" formation is best suited for fairly open terrain but can also be used in close terrain. When established in close terrain, the legs of the "V" close in as the lead element of the enemy force approaches the apex of the "V", and opens fire at a close range.

Here, even more than in open terrain, all movement and fire must be carefully coordinated and controlled to ensure that the fire of one wing does not endanger the other wing. The wider separation of forces makes this formation difficult to control, and there are few sites that favor its use. Its main advantage is that it is difficult for the enemy to detect the ambush until it is well into the killing zone. (See figs. D-9 and D-10.)

Triangle Formation

The triangle is a variation of the "V" formation and can be varied in three ways:

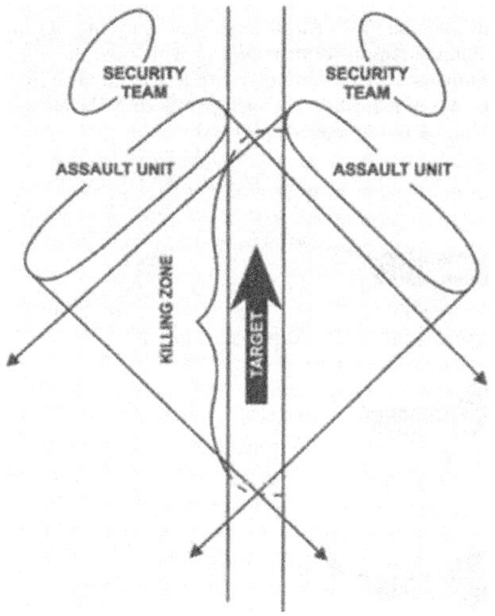

Figure D-9. "V" Formation (Open Terrain).

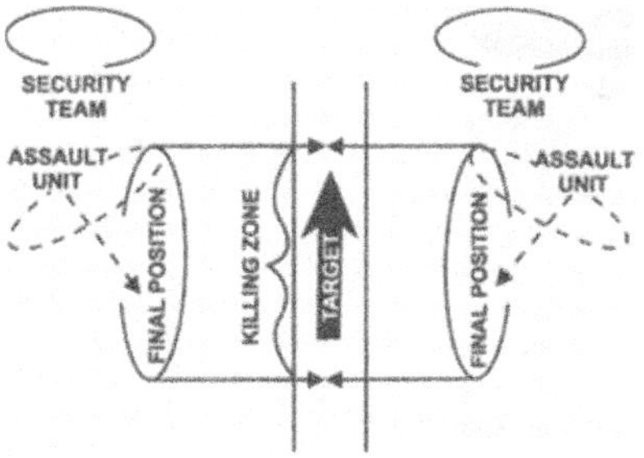

Figure D-10. "V" Formation (Close Terrain).

Closed Triangle Formation

(See fig. D-11.) The assault element is deployed in three teams, positioned so that they form a triangle (or closed "V"). An automatic weapon is placed at each point of the triangle and positioned so that it can be shifted quickly to interlock with either of the others. Men are positioned so that their sectors of fire overlap. Mortars may be positioned inside the triangle.

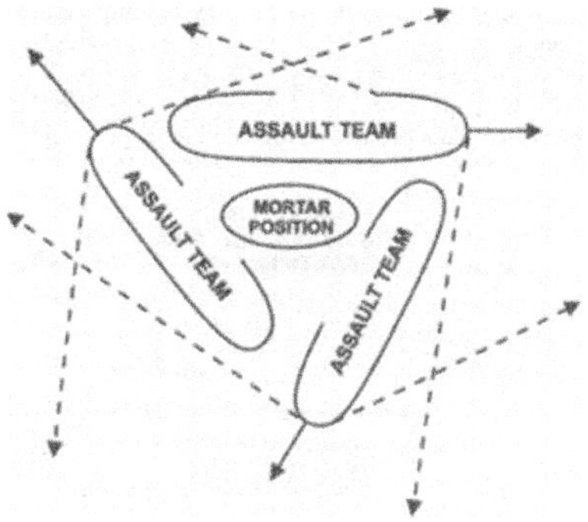

Figure D-11. Closed Triangle Formation (Night Ambush).

When deployed in this manner, the triangle ambush is used to interdict night movement through open areas. When enemy approach is likely to be from any direction, this formation provides all-around security, and security forces are deployed only when they can be positioned so that if detected by an approaching enemy, they will not compromise the ambush. Attack is by fire only, and the enemy is allowed to approach within close range before fire is initiated.

The advantages of the closed triangle formation are ease of control, all-around security, and the enemy can be brought under the fire of at least two automatic weapons, regardless of the direction they approach.

Disadvantages are that it requires a force of platoon size or larger to reduce the danger of being overrun by an unexpectedly large force and that one or more sides of the triangle may come under enfilade fire. The lack of dispersion, particularly at the points, increases the danger from enemy mortar fire.

Open Triangle Harassing Formation

This variation of the triangle ambush enables a small force to harass, slow, and inflict heavy casualties upon a large force without itself being decisively engaged. The assault element is deployed in three teams, positioned so that each team becomes a corner of a triangle containing the killing zone. (See fig. D-12.)

When the enemy enters the killing zone, the team to the enemy's front opens fire on the leading element. When the enemy counterattacks, the group withdraws and the team on the enemy's flank opens fire. When this team is attacked, the team to the opposite flank opens fire. This process is repeated until the enemy is pulled apart. Each team reoccupies its position, if possible, and continues to inflict the maximum damage possible without becoming decisively engaged.

Open Triangle Destruction Formation

The assault element is again deployed in three teams, positioned so that each team is a point of the triangle, 200 to 300 meters apart. The killing zone is the area within the triangle. The enemy is allowed to enter the killing zone; the nearest team attacks by fire. As the enemy attempts to maneuver or withdraw, the other teams open fire. One or more teams, as directed, assault or maneuver to envelop or destroy the enemy. (See fig. D-13 on page D-6.)

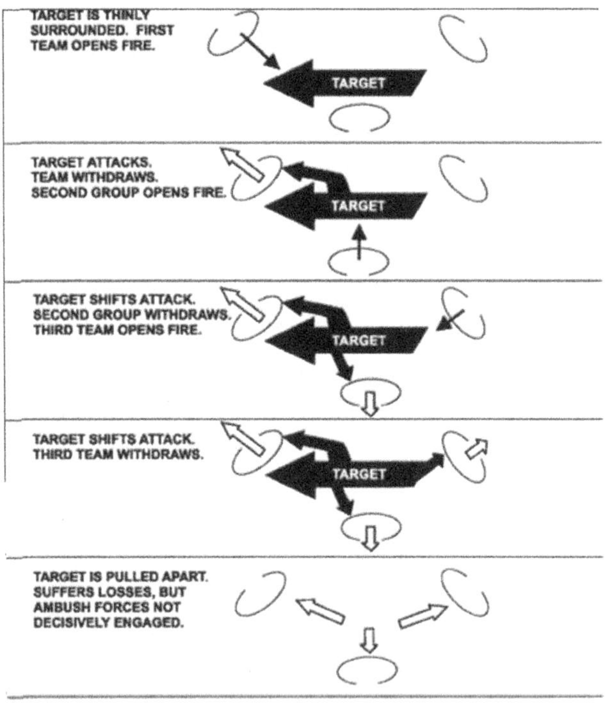

Figure D-12. Open Triangle Formation.

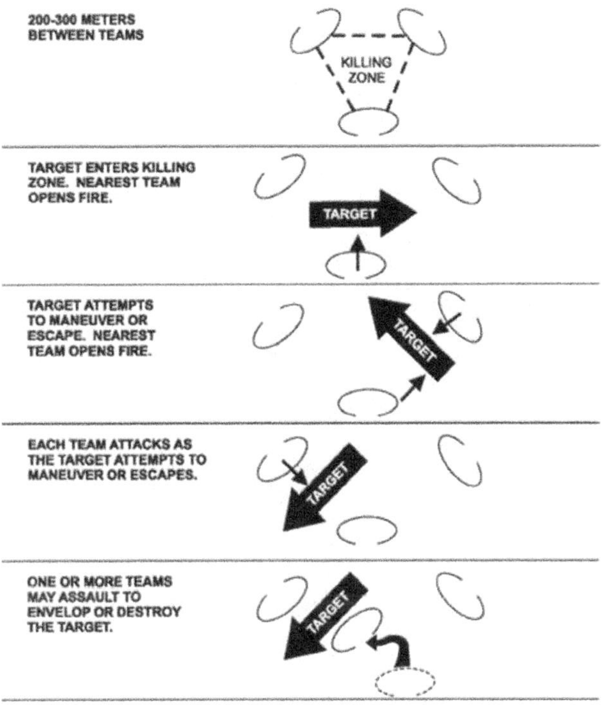

Figure D-13. Open Triangle Formation (Destruction Ambush).

This formation is suitable for platoon-size or larger ambush forces. A smaller force would be in too great a danger of being overrun. Another disadvantage is that control, in assaulting or maneuvering, is very difficult. Very close coordination and control are necessary to ensure that assaulting or maneuvering teams are not fired on by another team. The ambush site must be a fairly level open area that provides concealment around its border for the ambush force.

The Box Formation

The "box" formation is similar in purpose to the open triangle ambush. The assault element is deployed in four teams, positioned so that each team becomes a corner of a square or rectangle containing the killing zone. It can be used for a harassing or destruction ambush in the same manner as the variations of the open triangle formation. (See figs D-14 on page D-7 and D-15 on page D-8.)

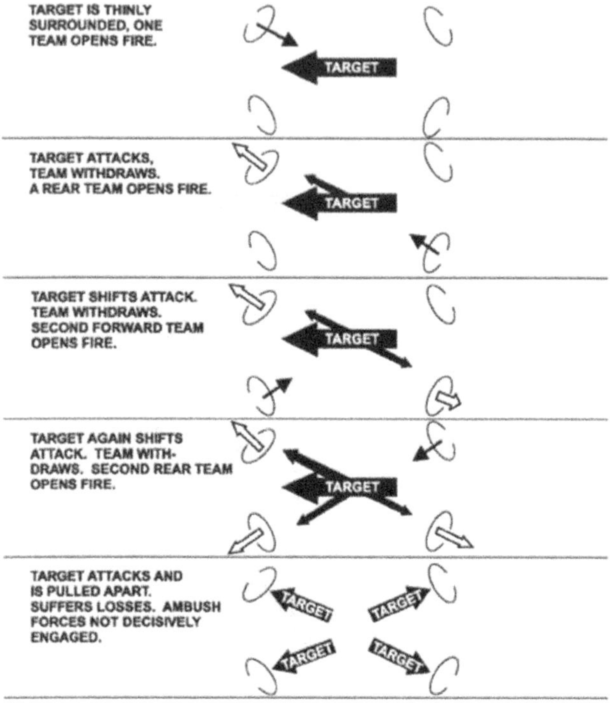

Figure D-14. Box Formation (Harrassing Ambush).

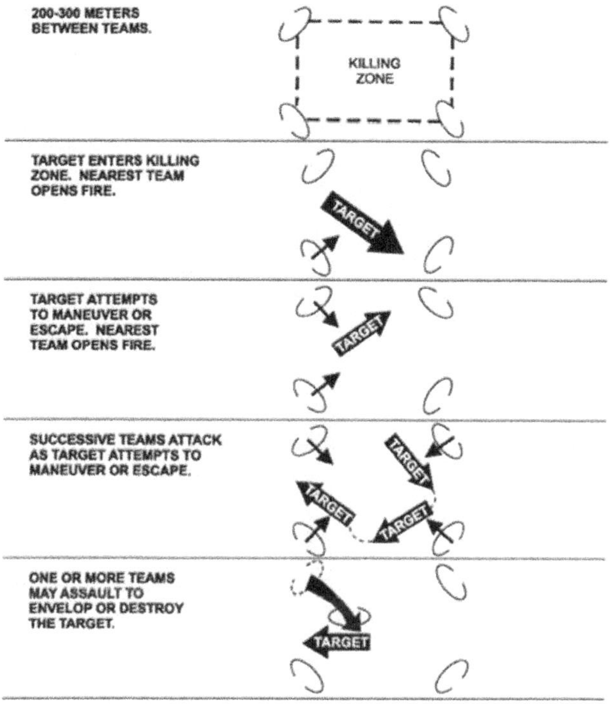

Figure D-15. Box Formation (Destruction Ambush).

APPENDIX E. ACRONYMS

A&S. . . . . assault and security

CIT . . . . counterintelligence team

EOD. . . . . explosive ordnance disposal

ITT. . . . . interrogator-translator team

MAGTF. . . . . Marine air-ground task force
MEDEVAC. . . . . medical evacuation
METT-T . . . . mission, enemy, terrain and weather, troops and support available

NBC. . . . . nuclear, biological, chemical
NCO . . . . noncommissioned officer

OTL. . . . . overserver-target line

RCA. . . . . riot control agents
ROE. . . . . rules of engagement

SMAW . . . . shoulder-launched multipurpose assault weapon
SMEAC. . . . . situation, mission, execution, administration
and logistics, and command and signal
SOP . . . . standing operating procedures
SPOTREP . . . . spot report

TTP . . . . tactics, techniques, and procedures

Appendix F. References

Joint Publication (JP)

1-02 DOD Dictionary of Military and Associated Terms

Fleet Marine Force Manual (FMFM)

6-5 Marine Rifle Squad (proposed MCWP 3-11.2)

Marine Corps Warfighting Publications (MCWP)

3-15.3 Scout Sniping (under development)

3-35.3 Military Operations on Urbanized Terrain (MOUT)

3-41.2 Raids (under development)

Marine Corps Reference Publications (MCRP)

2-15.3A Reconnaissance Patrol Leader's Planning Handbook (under development)

2-15.3B Reconnaissance Reports Guide

3-02C Water Survival Handbook (under development, currently exists as Fleet Marine Force Manual 0-13)

5-12A Operational Terms and Graphics

5-12C Marine Corps Supplement to the DOD Dictionary of Military and Associated Terms

Training Circular
No. 90-1

Headquarters
Department of the Army
Washington, DC, 19 May 2008

Training for Urban Operations

Contents

Distribution Restriction: Approved for public release; distribution is unlimited.

*This publication supersedes TC 90-1, 1 April 2002.

Figures

Tables

Chapter 1

Introduction

This chapter discusses definitions, training strategy, prerequisite training, individual task training, and collective task training for urban operations (UO). Military operations on urbanized terrain (MOUT) and UO differ in ways that reflect the changes in the way we engage in war today.

Section I. DEFINITIONS

This section defines and compares MOUT and UO.

> ***Note:*** The terms "urban operations" and "UO" replaces all instances of "military operations on urbanized terrain" and "MOUT."

MILITARY OPERATIONS ON URBANIZED TERRAIN

1-1. In the past, Army doctrine characterized urban areas as a type of specialized terrain, like jungles and mountains, and called operations in this type of terrain *military operations on urbanized terrain* (MOUT). Doctrine emphasized tactical warfighting tasks on this terrain, which was characterized by densely situated buildings and streets, and assumed the enemy consisted of conventional (Soviet-type) forces. Defeating or destroying those forces overrode all other considerations.

URBAN OPERATIONS

1-2. The concept of UO assumes that the urbanized terrain is populated, and that the populace must be a foremost consideration. UO recognizes that the populace can help either side, and that it requires a working infrastructure. Urban operations, whether against an insurgent or unconventional enemy, thus vastly differ from MOUT (Table 1-1).

Table 1-1. Comparison of MOUT and UO.

	MOUT	*UO*
Enemy force	Conventional, Soviet type	Insurgent or unconventional
Mission	Defeat or destroy enemy above all other considerations	Defeat the enemy Gain and retain the good will of the populace
Populace	Not considered	Important consideration

1-3. Although certain techniques and tactics are common to both types of operations, UO goes beyond MOUT, focusing not only on defeating the enemy, but also on gaining and retaining the good will of the population. UO consider the complex and dynamic interactions and relationships between the urban area's key components--the terrain (natural and manmade), the population, and the supporting infrastructure--as an overlapping and interdependent system of systems.

1-4. A single urban operation may include the full range of Army operations--offensive, defensive, stability and support (Appendix C)--that may be executed, either sequentially or simultaneously (FM 3-06, Chapter 1). Furthermore, to accomplish an objective, UO demand working with sister services, US government agencies, nongovernmental organizations, and regional and international organizations.

Section II. TRAINING STRATEGY

The UO training strategy is primarily organized by environment and phase. Training environments include live, virtual, and constructive (Figure 1-1), and are based on the unit's METL. Training phases are initial, intermediate, and culminating. Table 1-2 shows the relationship between the training environments and the first three training phases. (See also Appendix D, which discusses leader training.)

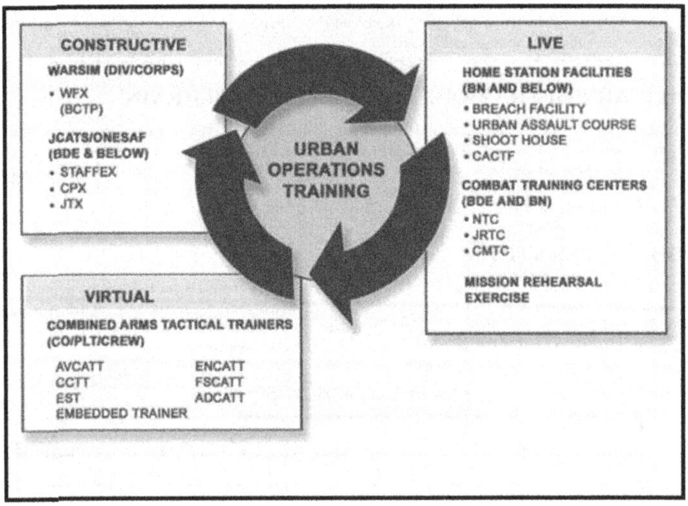

Figure 1-1. Urban Operations Training Support Strategy.

TRAINING PHASES

1-5. The three training phases are initial, intermediate, and culminating. Each of these includes some live, virtual, and constructive training.

INITIAL PHASE

1-6. This phase, which trains individual and collective tasks, is the first step in the building-block approach. It forms the basis of unit or collective training. For Soldiers and leaders, mastery of individual and crew skills is a prerequisite for UO training. The tasks train basic marksmanship skills and other Soldier tasks needed to perform any tactical operation. Included are training in a live environment (precision marksmanship) and virtual environment (Engagement Skills Trainer [EST], crew simulators, and CCTTs).

Table 1-2. Training environment/training phase matrix.

TRAINING PHASE / TRAINING ENVIRONMENT	INITIAL	INTERMED	CULMINATING
LIVE	Precision Marksmanship (Indiv / Sqd)	Breach Facility, Urban Assault Course, Shoot House (Team / Squad / Platoon)	CACTF, Platoon STX, Company STX, Battalion FTX Deployment CTC Rotation Mission Rehearsal Exercise
VIRTUAL	EST (Indiv / Sqd) Crew Simulators and CCTT (Crews Plt Ldrs, Co Cdrs)		
CONSTRUCTIVE			WARSIM (division/corps), which includes warfighting exercise (battle command training program). JCATS/OneSAF (brigade and below), staff exercises (STAFFEXs), command post exercises (CPXs), and joint training exercises (JTX)

INTERMEDIATE PHASE

1-7. This phase of training is characterized by collective task training in varied conditions such as limited visibility; different entry techniques and points; incorporation of combatants, noncombatants, and OPFOR; use of MILES; and live-fire exercises. It is the vital link between individual tasks and mission accomplishment. It should be conducted at, but is not limited to, the *Breach Facility*, *Live-Fire Shoot House*, and *UAC*. (Section III discusses these facilities in detail.) Many types of organizations can use these facilities. Trainers *must* be certified before conducting demolition training IAW local standing operating procedure (SOP) or ACOM policy. AR 385-63 prescribes safety requirements. Except for Aviation, Armor, and other branches that have no practical use for these facilities, this phase must be completed to advance to the culminating phase.

CULMINATING PHASE

1-8. The Combined Arms Collective Training Facility (CACTF, Figure 1-2) is where the culminating phase of urban operations training ends. This phase can include multiechelon, combined arms, and branch specific STX training may be conducted. Urban operations leader training occurs in leader professional development courses (Appendix D). Units have developed internal urban warfare training programs and TRADOC schools have assembled technical mobile training teams (MTTs) for marksmanship, combatives, and breaching techniques.

EDUCATION PHASE

- Urban warfare reading/video list.
- Urban operations equipment.
- Review of Infantry doctrinal publications (FM 3-21.10, FM 3-06.11, for Infantry), and corresponding publications for other branches.
- Small-unit urban movement techniques and considerations.

PLANNING AND EXECUTION PHASE

- Prepare company OPORD or WFF-related products.
- Brief platoon/company OPORD/WFF-related products to commander or staff representative.
- Participate as briefer in urban environment TEWT.

SUSTAINMENT PHASE

- Participate in company/platoon TEWT or other urban operations event.
- Execute/facilitate a company or platoon TEWT.

TRAINING ENVIRONMENTS

1-9. Both live and virtual training environments are used to train UO.

LIVE ENVIRONMENT

1-10. The live training environment consists of home station training using the Urban Assault Course (UAC), Shoot House (SH), Breach Facility (BF), Combined Arms Collective Training Facility (CACTF), Combat Training Center (CTC) rotations, and any other live training that supports the UO training.

VIRTUAL ENVIRONMENT

1-11. The virtual training environment consists of simulators such as Close Combat Tactical Trainer (CCTT) or Engagement Skills Trainer (EST), which are used to train small-unit leaders and crews. The constructive environment consists of battle staff training for battalion and above staff and leaders such as Joint Conflict and Tactical Simulation (JCATS)/OneSAF.

CONSTRUCTIVE ENVIRONMENT

1-12. The constructive training environment consists of the following, which all occur in the culminating training phase:

- WARSIM (division/corps), which includes warfighting exercise (battle command training program).
- JCATS/OneSAF (brigade and below), staff exercises (STAFFEXs), command post exercises (CPXs), and joint training exercises (JTX)

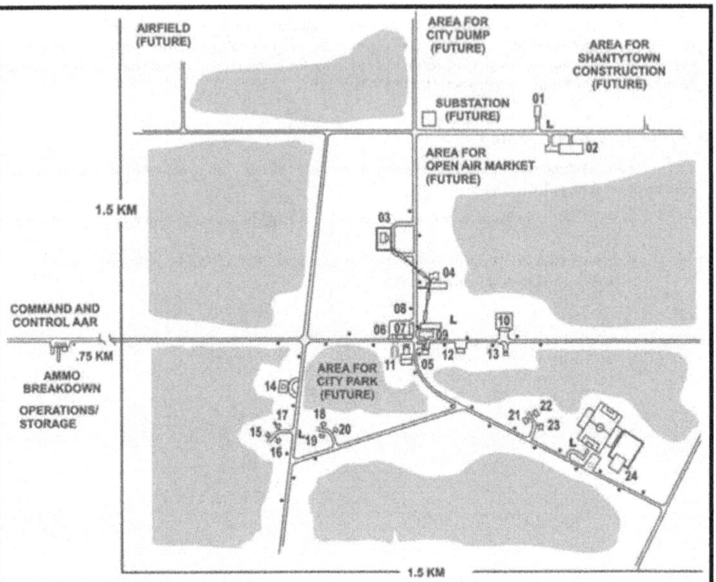

BLDG #	SIZE AND TYPE
01	1 STORY WAREHOUSE
02	1 STORY WAREHOUSE
03	3 STORY OFFICE (EMBASSY)
04	2 STORY OFFICE
05	1 STORY SERVICE STATION
06	1 STORY BUSINESS
07	1 STORY BUSINESS
08	1 STORY BUSINESS
09	3 STORY HOTEL
10	1 STORY POLICE STATION/JAIL
11	CHURCH AND CEMETARY
12	1 STORY BUSINESS
13	1 STORY BANK
14	2 STORY TOWNHOUSE
15	1 STORY RESIDENT
16	1 STORY RESIDENT
17	1 STORY RESIDENT
18	1 STORY RESIDENT
19	1 STORY RESIDENT
20	1 STORY RESIDENT
21	1 STORY RESIDENT
22	1 STORY RESIDENT
23	1 STORY RESIDENT
24	2 STORY SCHOOL

LEGEND

- FENCE WALL
- WALL
- TUNNEL SYSTEM
- TREE LINE
- UTILITY POLE
- LATRINE

SCALE: 1" = 500'

Figure 1-2. Combined Arms Collective Training Facility.

MISSION-ESSENTIAL TASK LIST

1-13. FMs 7-0 and 7-1 explain how to determine a unit's mission-essential task list (METL) and battle tasks, and how to plan, resource, and execute training events. Because training facilities are limited, leaders must use all available time and resources for training and maintaining combat readiness.

- System-specific FMs and TCs teach doctrine, conduct of operations, and employment of weapon systems.
- Tactics, techniques, and procedures (TTP) are lessons learned from training and combat.

1-14. The standards for evaluating training are accessible at the Digital Training Management System (DTMS), https://dtms.army.mil/DTMS.

BATTALION INTEGRATED TRAINING PLAN

1-15. The commander uses the battalion integrated training plan to set goals for training density. He focuses on combined arms operations, emphasizing dismounted Infantry maneuver. Tanks, BFVs, Strykers, attack aviation, and artillery normally provide supporting fires for the dismounted elements. UO training in the CACTF, Shoot House, and Breach Facility should be conducted semiannually and quarterly for the UAC. Army training calendars are divided into two cycles of training: *mission* and *support*. Figure 1-3 and Figure 1-4 show how the urban training strategy and facility usage might be programmed into unit cycles.

RISK ASSESSMENT

1-16. Risk assessment means making operations safe without compromising the level of realistic training (FM 7-1). Commanders continuously assess the risk of training conditions to prevent unnecessary loss of Soldiers and equipment. The degree of risk depends on conditions. Just as this principle applies to other training and warfighting, it also applies to UO. In a training situation, for example, commanders determine whether Soldiers and leaders have conducted this type of training before, whether conducting initial training at night or during limited visibility, and whether Soldiers are tired or exhausted. Risk assessment ensures UO training is not only tough and realistic but also safe. Training enhancers, such as live fire, blank ammunition, simulators, fires, smoke, and other pyrotechnics, must be fully integrated into the training to achieve the training objectives. Soldiers are required to wear earplugs, protective vests, and protective eyewear in order to protect themselves from noise and debris.

1-17. The commander considers risk and safety in urban operations training situations such as--.

- Soldiers engaging in close combat urban operations need to be safety conscious when using blanks, booby traps, and practice grenades. They may be unaware of the exact location of the OPFOR and may be susceptible to OPFOR flash suppressors, booby traps, and practice hand grenades.
- Soldiers should use ladders to enter buildings at upper stories of buildings to prevent unnecessary risks.

1-18. The commander is the safety officer and is responsible for ensuring that his unit, both leaders and Soldiers, comply with safety regulations and the units tactical SOP. All leaders must--.

- Use mission, enemy, terrain, troops, time available, and civilians (METT-TC) factors to identify risks.
- Assess possible losses and their costs.
- Select and develop risk-reduction measures.
- Implement controls by integrating them into plans and orders, SOPs, training performance standards, and rehearsals.
- Supervise and enforce risk reduction measures and safety standards at all times.

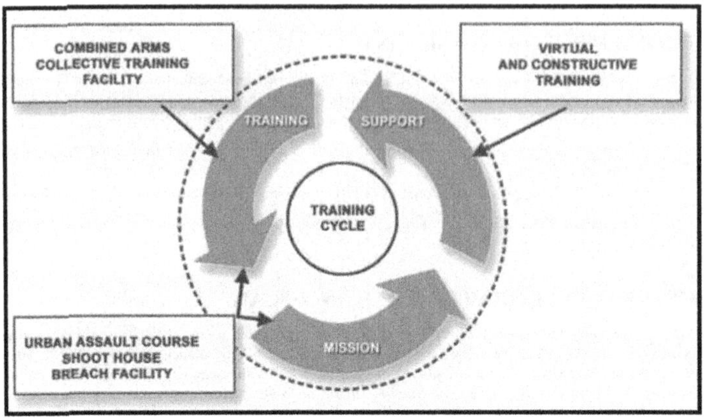

Figure 1-3. Active Army, Home Station, Urban Operations Training Strategy.

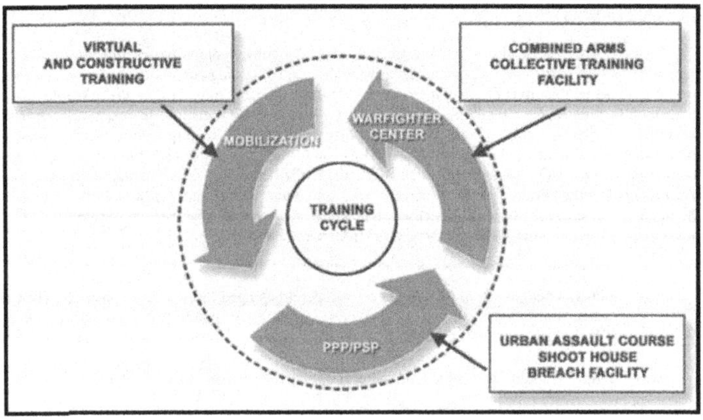

Figure 1-4. Army National Guard Urban Operations Training Strategy.

Section III. TRAINING FACILITIES

Training facilities include three that are discussed in this section: the Urban Assault Course (UAC, Chapter 2), Shoot House (SH, Chapter 3), Breach Facility (BF, Chapter 4). They also include the Combined Arms Collective Training Facility (CACTF, discussed in Chapter 5), Combat Training Centers (CTC), and any other live training that supports the UO training.

URBAN ASSAULT COURSE

1-19. The purpose of the UAC facility is to train a squad-sized force to *Enter a Building and Clear a Room*, improve grenadier gunnery, and validate subterranean TTP. However, the unit's training requirements determine the size of the unit as well as the tasks to be trained. The unit develops its organizational structure and steps for conducting the exercise based on its METL and on their training needs (Figure 1-5). (See Chapter 2 for a more detailed discussion.)

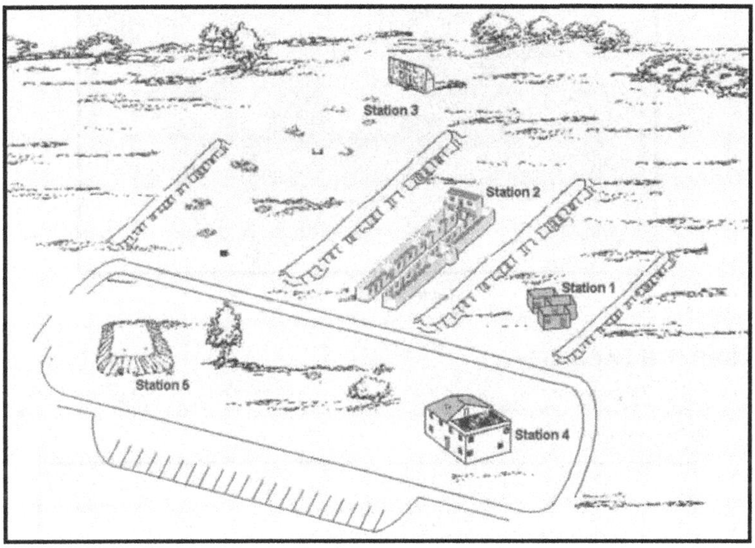

Figure 1-5. Urban Assault Course.

SHOOT HOUSE

1-20. The purpose of this facility is to provide the leader with a facility to train and evaluate fire teams, squads, and platoons during a live-fire exercise (Figure 1-6). Units are trained and evaluated on their ability to conduct tactical movement and engage targets, while practicing target discrimination in an urban environment. (See Chapter 3 for a more detailed discussion.)

Note: The specific designs of the Shoot House and the other facilities may differ from one installation to another, due to variations in training requirements.

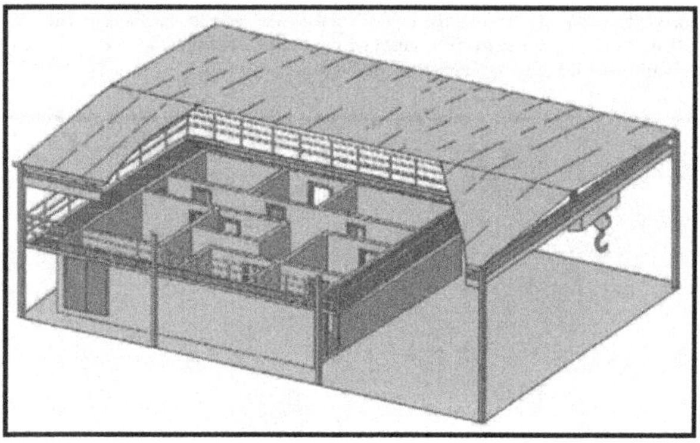

Figure 1-6. Shoot House.

BREACH FACILITY

1-21. The purpose of this facility is to train Warfighters on the technical aspects of mechanical, ballistic, thermal, and explosive breaching techniques (Figure 1-7), including for use in UO. (See Chapter 4 for a more detailed discussion.)

Figure 1-7. Breach Facility.

Section IV. CONDUCT OF TRAINING

Depending on unit proficiency, every training event should follow a *crawl-walk-run* sequence, based on clearly defined task, conditions, and standards. After each training event, leaders conduct an after-action review (AAR).

CRAWL

1-22. The trainer explains each training objective and performance standards, and then talks the Soldiers through the exercise, step by step, describing what each individual or unit must do. This training often includes a demonstration.

WALK

1-23. The Soldiers slowly practice each task to standard. Trainers coach Soldiers at this stage stopping as often as needed to correct mistakes and provide feedback. Soldiers practice each task repeatedly until they can perform it to standard.

RUN

1-24. Soldiers perform each task at full speed as if they are in combat. The Soldiers need feedback during this stage as well. The following can be used to enhance realism:

- Blank ammunition.
- Special Effects Small-Arms Marking System (SESAMS).
- Targetry.
- Pyrotechnics.
- Tactical engagement systems (MILES).
- Opposing forces (force on force).
- Nuclear, biological, and chemical simulation/weapons of mass destruction.
- Training in limited visibility conditions.

AFTER-ACTION REVIEW

1-25. Leaders conduct an after-action review (AAR) after each training event to help Soldiers and units improve their skills by providing immediate feedback. An AAR increases the benefits gained from each training exercise by allowing the leaders and the Soldiers to work together to analyze the performance of each task. Being involved in this analysis enhances the leaders and the Soldiers learning and retention ability. The AAR improves the performance of the leaders and Soldiers in later iterations by allowing interaction and feedback from these individuals during the debrief.

Section V. PREREQUISITE TRAINING

Prerequisite training teaches individuals, crews, units, leaders, and command and staff what they need to know before they are required to perform a task. Prerequisite training builds teamwork and individual confidence at all levels.

COMMAND AND STAFF TRAINING

1-26. The battalion battle staff and leadership must train before the battalion is required to conduct collective training. The battalion command and staff culminates their prerequisite training with a tactical exercise without troops (TEWTs) and command post exercises (CPXs). This sequence of training assists the commander and his staff develop better UO TTP and unit SOP.

LEADER TRAINING

1-27. A comprehensive leader's training program is the key to successful UO training (Appendix D). Time spent teaching and training the battalion leaders (trainers) during the planning and preparation phase will ensure success during the intermediate and culminating phases of training. Realistic training at the unit level will improve the leadership training officers and NCOs receive at the Basic Officer Leadership Course (BOLC), Captains Career Officer Course (CCOC), Basic Noncommissioned Officer Course [BNCOC], and Advanced Noncommissioned Officer Course [ANCOC]. Recommended training should include remedial training on those tasks that the unit had difficulty in during the previous training and evaluation.

INDIVIDUAL TRAINING

1-28. Many individual specific tasks and some collective tasks can be taught and practiced in garrison while most collective tasks are best conducted at the installation UO facilities and weapons ranges. The unit's preliminary individual training tasks should focus on three general areas: urban combat skills, urban marksmanship skills, and physical fitness. The following are example preliminary training tasks or TTP:

- Individual movement techniques.
- Battle drills.
- Weapon positioning in an urban area.
- Roadblock and vehicle search procedures.
- Dismounted and mounted urban navigation.
- Urban scanning, quick-fire, and assault fire techniques.
- Hasty urban firing positions.
- Preparation of urban firing positions.

1-29. Prior to urban collective training cycle, trainers should develop urban-specific firing conditions in conjunction with weapons firing ranges. Build facades on existing ranges, or use the unit's local UAC (if available). Urban combat skills include--

- Firing positions behind rubble, around corners, from windows or rooftops.
- Urban targets behind rubble, in windows and doorways.
- Fire commands.
- Lifting and shifting fires from lower to upper story windows.
- Fire control and fire discipline.
- Rapid firing engagements.
- ROE firing scenarios.

1-30. Urban physical fitness skills include--

- Endurance runs and upper body conditioning.
- Vaulting exercises.

- Urban-specific obstacle course negotiation.
- Leadership Reaction Course negotiation.

INDIVIDUAL TASK TRAINING

1-31. Urban operations require the Soldier to be proficient in those individual tasks that are unique to urban combat. Units should incorporate into their training regime those urban fighting TTP that have proven successful based on lessons learned from combat operations.

TASK FORCE SOLDIER--WARRIOR COMMON TASKS, SKILL LEVELS 1-3:

- 071-326-0541, *Perform Movement Techniques during an Urban Operation.*
- 071-326-0542, *Enter a Building during an Urban Operation.*
- 071-326-0550, *Prepare Positions for Individual and Crew-Served Weapons during an Urban Operation.*
- 071-326-0557, *Select Hasty Firing Positions during an Urban Operation.*
- 071-440-0027, *Execute Mechanical Breach in an Urban Operation.*
- 071-440-0028, *Engage Targets during an Urban Operation.*
- 071-440-0029, *Conduct a Breach during an Urban Operation.*
- 071-440-0003, *Conduct an Attack on a Building by a Squad during an Urban Operation.*
- 071-440-0006, *Conduct a Defense by Squad during an Urban Operation.*

FM 3-06.11, *COMBINED ARMS OPERATIONS IN URBAN TERRAIN*

Movement Techniques

- Crossing of a wall.
- Movement around corners.
- Movement past windows.
- Use of doorways.
- Movement parallel to buildings.
- Crossing open areas.
- Fire team employment.
- Movement between positions.
- Movement inside a building.

Entry Techniques

- Upper building levels.
- Use of ladders.
- Scaling of walls.
- Entry at lower levels.
- Hand grenades.

Firing Positions

- Hasty firing position.
- Prepared firing position.
- Target acquisition.
- Employment of snipers.

COLLECTIVE TASK TRAINING

1-32. Specific FMs teach doctrine, how to conduct operations, and how to employ weapon systems. TTP are lessons learned from the training and combat. The standards for conducting and evaluating collective training are accessible at the Digital Training Management System (DTMS), https://dtms.army.mil/DTMS. Those tasks that do not specifically address UO must be modified for use in an urban environment.

RESOURCES

1-33. Field manuals for the squad and platoon are--:

- FM 3-21.8, *Infantry Rifle Platoon and Squad.*
- FM 3-21.71, *Mechanized Infantry Platoon and Squad (BFV).*
- FM 3-21.9, *SBCT Infantry Rifle Platoon and Squad.*

1-34. Field manuals for the company are--.

- FM 3-21.10, *IBCT Rifle Company.*
- FM 3-21.11, *SBCT Rifle Company.*

1-35. Field manuals and mission training plans for the battalion are--.

- FM 3-21.20, *IBCT Infantry Battalion.*
- FM 3-21.21, *SBCT Infantry Battalion.*
- FM 3-06.11, *Combined Arms Operations in Urban Terrain.*
- TC 7-98-1, *Brigade and Battalion Stability and Support Operations.*

Chapter 2
Urban Assault Course

This chapter describes the Urban Assault Course (UAC), which incorporates doctrine and tactics from FM 3-06.11. The UAC contains five stations:

- Individual and Team Task/Technique Trainer Station 1
- Squad and Platoon Task/Technique Trainer Station 2
- Grenadier Gunnery Trainer Station 3
- Urban Offense/Defense Building Station 4
- Underground Trainer Station 5

These stations are designed for individual tasks and small-unit collective task training. Various types of organizations with a variety of missions can train at the UAC. Tactical operations in an urban environment may include combat to stability and support tasks. (Appendix C provides a SASO unit training matrix). The UAC is a dismounted training facility.

Note: The station discussions are arranged so that they may be removed from the book for ease of use.

PURPOSE

2-1. The purpose is to provide squad and platoon size units with a facility to train and evaluate urban operations tasks (Figure 2-1). Only Station 3, Grenadier Gunnery, is intended for live-fire training. It supports use of 40-mm TP and 5.56-mm service ammunition.

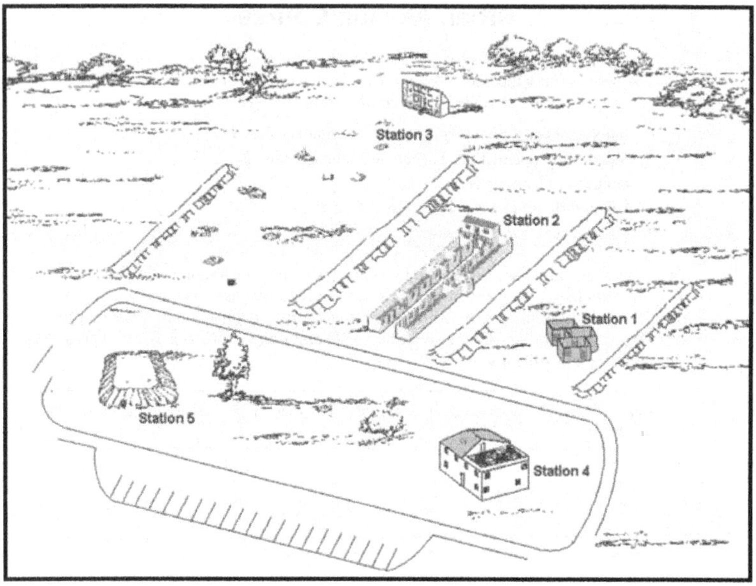

Figure 2-1. Urban Assault Course.

DESCRIPTION

2-2. The UAC has five training stations that facilitate the ***crawl-walk-run*** training concept:

Station 1, Individual and Team Trainer

2-3. This station has three rooms where team leaders and squad leaders train the basics of building and room clearing.

Station 2, Squad and Platoon Trainer

2-4. This station has four structures and multiple rooms. Squads build on tasks learned at Station 1, and begin to learn the concepts of clearing multiple buildings. This station can be used as individual buildings with a narrow street, or as rooms inside a building with a long, connecting hallway.

Station 3, Grenadier Gunnery Trainer

2-5. This is a live-fire station where M203 gunners master target engagements in an urban area, conduct tactical movement, and respond to the fire commands.

Station 4, Offense/Defense House

2-6. This is where a platoon trains to attack or defend a building. This station can also be divided into smaller training stations to reinforce training. It is also used to train tasks not yet trained at the other stations such as upper-level entry techniques.

Station 5, Underground Trainer

2-7. This station provides training for subterranean operations.

TARGETRY

2-8. Targets throughout the UAC are either 3D precision targets (for engagements less than 50 meters) or 2D nonprecision targets (for engagements greater than 50 meters).

Precision Targets

2-9. Precision targets support short distance engagement sensors. These sensors, located inside the precision targets, respond only to lethal shot placements.

- Torso lethal zone .... 18" wide x 8" high
- Neck-head lethal zone .... 4" wide x 8" high

Nonprecision Targets

2-10. Nonprecision targets support engagements at ranges greater than 50 meters. These targets are 2D pop-up type targets.

Training Audience

2-11. The facility gives Infantry, Engineer, Military Police, Cavalry, and other units a facility for training individual Soldiers through platoon-level UO skills.

Training Information

2-12. Combat units can use this facility as a mission rehearsal exercise site. Recommended training frequency is quarterly. For each UAC station, they should provide the purpose, a description, targetry, controller and safety requirements, individual UO tasks to be trained, training support requirements, and conduct of training information, if applicable:

Purpose

2-13. This UAC station [*state purpose of station*]. [*Units may use the station in the manner originally intended, or they may modify it to meet their unique training needs.*]

Description

2-14. This station is [*describe station use and design*].

Targetry

2-15. Recommended targetry for training at this station includes [*describe*]. Targets may be moved, modified, or added, depending on available target systems or training requirements.

Controller

2-16. Controller requirements for this station include [*describe*].

Safety

2-17. Safety requirements and considerations for this station include [*describe*]. [*The unit or installation range SOP may add more stringent safety requirements.*]

Individual Tasks or Skills Trained

2-18. These may be found in the Digital Training Management System located online at https://dtms.army.mil/DTMS.

Training Support

2-19. Minimum materiel requirements to support training at this station include [describe].

Conduct of Training

2-20. Example training scenarios and sequence of events for this station include [describe]. [The UAC can accommodate an assortment of training ammunition and pyrotechnics.]

STATION 1: INDIVIDUAL AND TEAM TASK/TECHNIQUE TRAINER

2-21. This station is used to train individual and collective tasks, tactics, techniques, and procedures, specifically *Enter a Building and Clear a Room* and *Engage Targets* (Figure 2-2). It has three rooms where team leaders and squad leaders train the basics of building and room clearing.

DESCRIPTION

2-22. Station 1 has a single wood structure. Its three rooms have one or more open doorways and windows, and a man-sized hole.

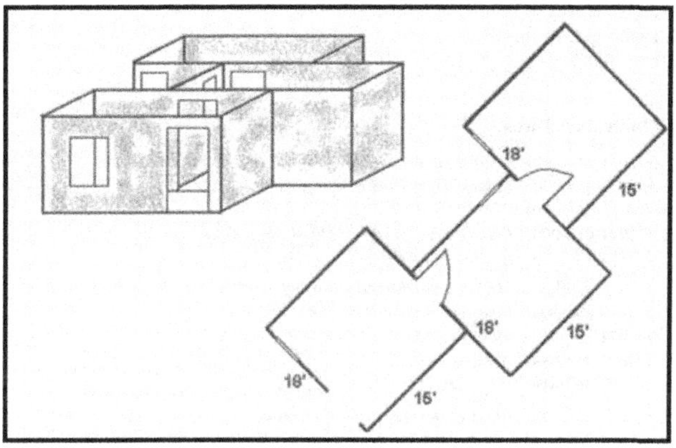

Figure 2-2. Individual and team task/technique trainer, Station 1.

TARGETRY

2-23. Targets present realistic close quarters combat situations (0 to 15 meters), which consist of a possible mixture of combatant and noncombatant targets. Target discrimination is trained by presenting up to six precision targets. Targetry sensors respond only to lethal shot placement. They also aid in training during limited visibility.

CONTROLLER REQUIREMENTS

2-24. These are based on unit requirements.

SAFETY REQUIREMENTS

2-25. Only smoke and practice grenades, or M84 stun grenades, may be used at this station.

> **WARNING**
>
> **To avoid injury, place the weapon on SAFE when climbing through obstacles such as windows, fences, and walls during training.**

INDIVIDUAL/COLLECTIVE TASKS TRAINED

2-26. This is noninclusive.

Collective Task

2-27. *Enter a Building/Clear a Room.*

Supporting Individual Tasks

- 071-311-2007, *Engage Targets with M16-Series Rifle/M4 Carbine Rifle.*
- 071-315-2308, *Engage Targets with M16-Series Rifle/M4 Carbine Using a Night Vision Sight AN/PVS-4.*
- 071-010-0006, *Engage Targets with M249 Machine Gun.*
- 071-325-4407, *Employ Hand Grenades.*
- 071-326-0503, *Move Over, Through, or Around Obstacles (except Minefields).*
- 052-193-1013, *Neutralize Booby Traps.*
- 071-326-0501, *Move as a Member of a Fire Team.*
- 071-326-0541, *Perform Movement Techniques during an Urban Operation.*
- 071-326-0557, *Select Hasty Firing Positions during an Urban Operation.*
- 071-326-5605, *Control Movement of a Fire Team.*
- 071-326-5611, *Conduct the Maneuver of a Squad.*
- 071-710-0008, *Operate Night Vision Goggles AN/PVS-7B.*

Related Tactics, Techniques, and Procedures

- FM 3-06.11, *Combined Arms Operations in Urban Terrain.*
- FM 3-22.9, *Rifle Marksmanship.*

TRAINING SUPPORT REQUIREMENTS

2-28. Use Table 2-1 for the minimum ammunition required to train at this station.

Table 2-1. Support requirements for Station 1.

Ammunition	*Rounds per Weapon*
5.56-mm Blank	10 rounds
5.56 mm Blank (Linked)	15 rounds
Practice Grenade Bodies	2 per team
Practice Grenade Fuses	2 per team

Conduct of Training

2-29. Table 2-2 lists the training sequence and times for Station 1.

Table 2-2. Sequence of events and times for Station 1.

Event	*Action*	*Estimated Time*
1	*Employ Hand Grenades*	30 minutes crawl, walk, run
	Perform Movement Techniques • *Engage Target with M16* • *Target Discrimination*	20 minutes crawl 20 minutes walk 10 minutes run
2	*Enter a Building*	20 minutes crawl 20 minutes walk 10 minutes run
3	*Enter and Clear a Room*	20 minutes crawl 20 minutes walk 10 minutes run
4	*Conduct an AAR*	40 minutes

Example Scenarios, Assault Team, Varied Entry Positions

2-30. Figures 2-3 and 2-4 show example scenarios. Rather than using the examples, units should tailor their training events based on their unit's METL and training needs.

Varied Entry Positions

2-31. The team leader first trains to execute *Varied Entry Positions* without targets, which is part of the *crawl* phase of training. The trainer has the assault teams position at various places to demonstrate how the start position can affect the final points of domination. (Figure 2-3).

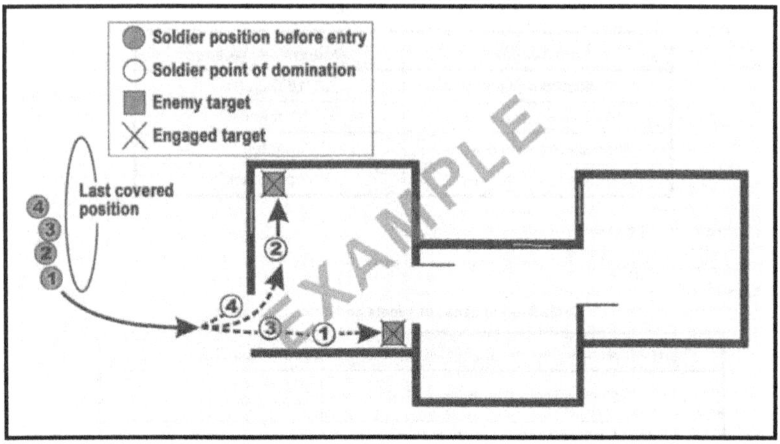

Figure 2-3. Example Station 1 scenario, *Varied Entry Positions*.

Multiple Room-Clearing Techniques

2-32. The team leader trains his Soldiers to execute *Varied Entry Positions* using targets in a multiple room scenario. This is part of the *walk* phase. The trainer has the assault teams enter the room and engage targets. He has them use using precision marksmanship techniques with the Tactical Engagement System (TES) or SESAMS (Figure 2-4). The assault team enters the first room and engages the targets. As the number two Soldier clears his area of responsibility, he sees and engages the target in the second room, but remains in place until the assault teams finishes clearing the first room. Once the room is clear, the assault team moves to the second room and continues clearing, or lets the next team move through.

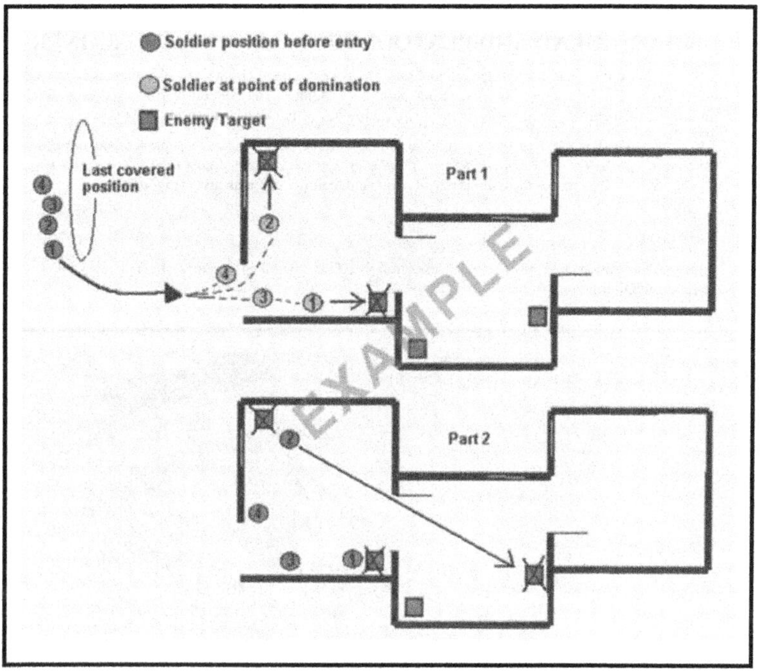

Figure 2-4. Example Station 1 scenario,
Multiple Room-Clearing Techniques.

STATION 2: SQUAD AND PLATOON TASK/TECHNIQUE TRAINER

2-33. This station has four structures and multiple rooms. Squads build on tasks learned at Station 1, and begin to learn the concepts of clearing multiple buildings. This station can be used as individual buildings with a narrow street, or as rooms inside a building with a long, connecting hallway.

2-34. Station 2 is used to train individual and collective tasks, tactics, techniques, and procedures, specifically, *Move Tactically Down a Street (Hallway)*, *Enter a Building and Clear a Room*, and *Engage Targets*. Station 2 increases the complexity of command, control, and maneuver (Figure 2-5).

DESCRIPTION

2-35. This station has four separate structures that face inward to create a street. It also introduces a two-story building.

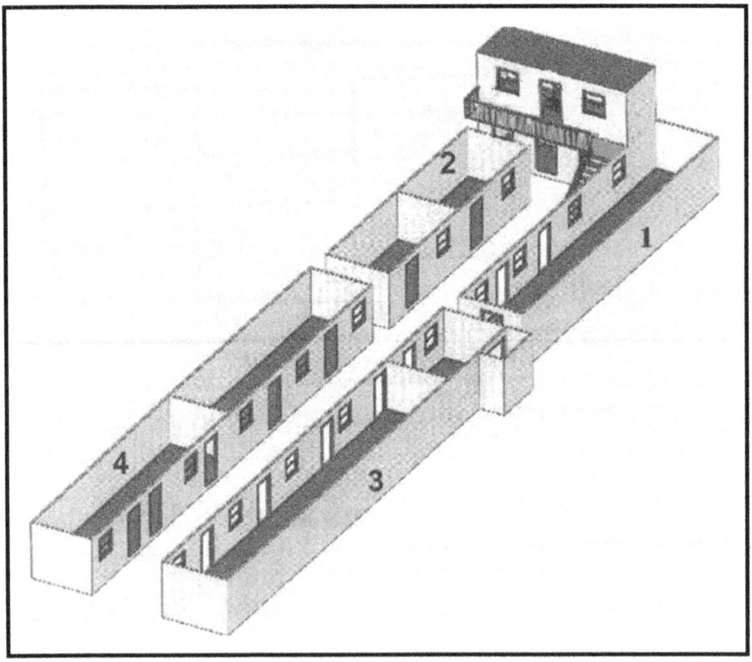

Figure 2-5. Squad and platoon task and technique trainer, Station 2.

Building 1 is L-shaped.

- The long leg of the building to the right looks like a single-story building.
- The short leg of the building to the front has two floors. A landing along the length of the second floor has outside stairs. An inside stairway also joins the floors.
- The two legs of the "L" are adjacent, but without interior access to each other.

Building 2 has a single floor and two rooms, each with a door and window. An open area much like a street separates this structure from the first.

Building 3 is a row of wooden buildings along a street. It has many windows and doorways. At the far end is an adjoining room.

Building 4 is on the left. It mirrors Building 3, plus it has a wall on the side at the near end.

2-36. Units can place additional walls in any structure by using target cloth or other material. They can do this to add to the number of rooms or to increase the complexity of the training (Appendix A).

TARGETRY

2-37. Station 2 targets present realistic close quarters combat situations (0 to 15 meters), with the option to customize the mixture of combatant and noncombatant targets. Target discrimination is trained on up to ten precision targets. Targetry sensors respond only to lethal shot placement, and facilitate training during limited visibility.

CONTROLLER REQUIREMENTS

2-38. These are based on unit requirements.

SAFETY REQUIREMENTS

2-39. At this station, Soldiers may only use smoke grenades, practice grenades, and M84 stun grenades.

WARNING

To avoid injury, place the weapon on SAFE when climbing through obstacles such as windows, fences, and walls during training.

INDIVIDUAL AND COLLECTIVE TASKS TRAINED

2-40. This list is noninclusive.

Collective Task--*Enter and Clear a Building*

Related Collective Tasks

- *Conduct Consolidation and Reorganization (Company/Platoon).*
- *Treat Casualties.*
- *Evacuate Casualties.*
- *Defend MOUT Building.*

- *Conduct Tactical Movement.*
- *Enter and Clear a Room.*

Supporting Individual Tasks

- 052-193-1013, *Neutralize Booby Traps.*
- 071-010-0006, *Engage Targets with M249 Machine Gun.*
- 071-311-2007, *Engage Targets with M16-Series Rifle/M4 Carbine Rifle.*
- 071-315-2308, *Engage Targets with M16-Series Rifle/M4 Carbine Using a Night Vision Sight AN/PVS-4.*
- 071-325-4407, *Employ Hand Grenades.*
- 071-326-0501, *Move as a Member of a Fire Team.*
- 071-326-0503, *Move Over, Through, or Around Obstacles (except Minefields).*
- 071-326-0541, *Perform Movement Techniques during an Urban Operation.*
- 071-326-0557, *Select Hasty Firing Positions during an Urban Operation.*
- 071-326-0608, *Use Visual Signaling Techniques.*
- 071-326-5605, *Control Movement of a Fire Team.*
- 071-326-5611, *Conduct the Maneuver of a Squad.*
- 071-710-0008, *Operate Night Vision Goggles AN/PVS-7B.*
- 113-571-1022, *Perform Voice Communications.*
- 191-377-5250, *Handle Enemy Personnel and Equipment.*

Related Tactics, Techniques, and Procedures

- FM 3-06.11, *Combined Arms Operations in Urban Terrain.*
  -- Chapter 3, *Offensive Operation.*
  -- Chapter 5, *Fundamental Combat Skills.*
  -- Appendix M, *Field-Expedient Breaching of Common Urban Barriers.*
- FM 3-22.9, *Rifle Marksmanship.*

TRAINING SUPPORT REQUIREMENTS

2-41. Table 2-3 shows the minimum ammunition required to train at this station.

Table 2-3. Support requirements for Station 2.

Ammunition	*Rounds per Weapon*
5.56-mm blank	16
5.56-mm blank (linked)	24
Practice grenade fuses	8 per squad
Practice grenade bodies	8

Conduct of Training

2-42. Table 2-4 shows the sequence of events and times for Station 2. Figure 2-6 through Figure 2-28 show example scenarios. However, rather than use these examples, the unit should customize its training events based on its METL and training needs.

Table 2-4. Sequence of events and times for Station 2.

Event	*Action*	*Estimated Time*
1	*Conduct Tactical Movement in a Built-Up Area* *Move Across an Open Area* *Move Parallel to Buildings*	10 min Crawl 20 min Walk 10 min Run
2	*Enter and Clear a Room (Squad)* *Clear a Stairwell* *Clear a Hallway* *Conduct a Breach*	10 min Crawl 20 min Walk 10 min Run
3	*Enter and Clear a Building (Platoon)* *Clear Multiple Rooms* *Consolidate and Reorganize*	10 min Crawl 20 min Walk 10 min Run
4	*Conduct AAR (Squad)*	10 min
NOTE: Training should include techniques outlined FM 3-06.11.		

2-43. The squad is clearing Station 2 as if it were a single building, and the street a hallway. Fire Team A moves down the left side, throws the appropriate grenade IAW ROE, enters, engages the target, and clears the room on that side. The last Soldier remains outside and secures the rear Fire Team B prepares to enter the next room (Figure 2-6).

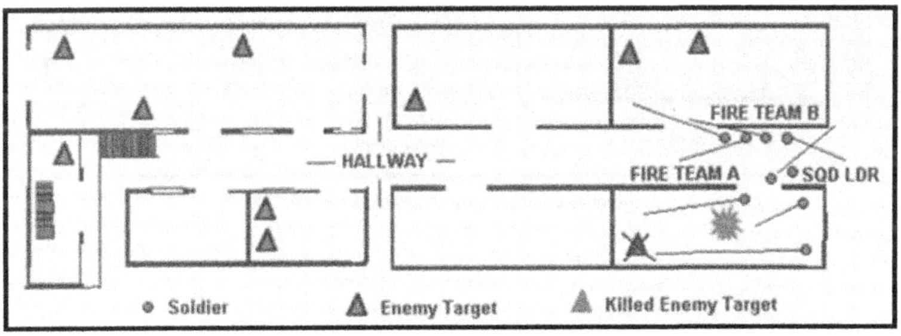

Figure 2-6. Conducting tactical movement.

2-44. Fire Team B throws the appropriate grenade, enters, engages the targets, and clears the room on the right while the last Soldier secures the rear Fire Team A prepares to continue movement to the next room (Figure 2-7).

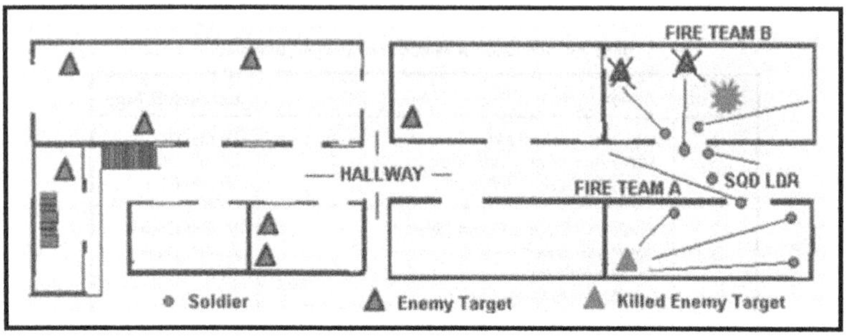

Figure 2-7. Team B--Clearing a building.

2-45. Fire Team A moves across the hall, throws the appropriate grenade, enters, engages the target, and clears the next room Three team members enter the room, the fourth remains in the doorway and secures the hallway (Figure 2-8).

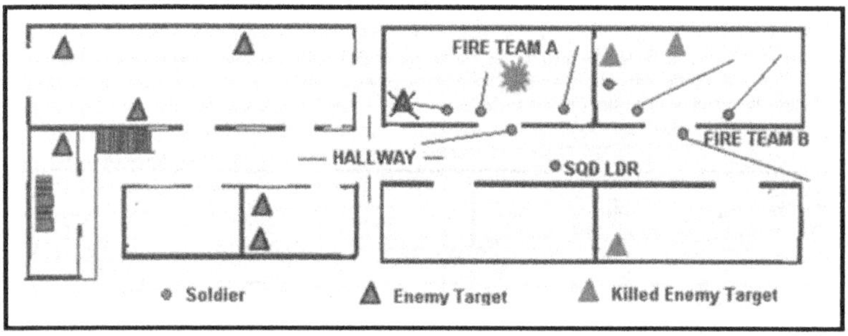

Figure 2-8. Team moving across the hall.

2-46. Fire Team B moves in a cross-cover formation down the hallway to the next room, throws the appropriate grenade, and clears the room. As Fire Team B passes the security man of Team A, he turns and becomes rear security. The last Soldier in Fire Team B remains by the door and secures the hallway (Figure 2-9).

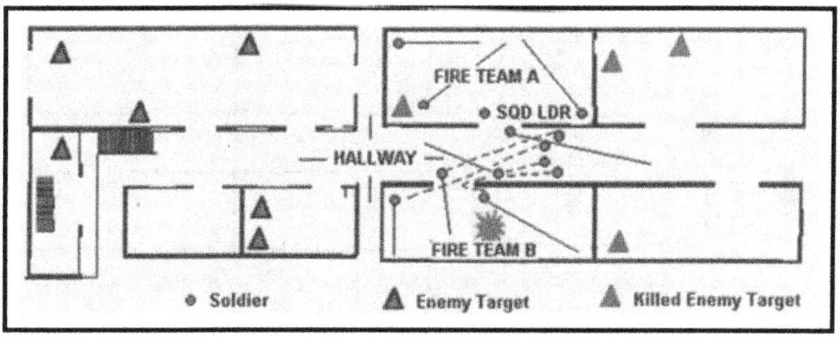

Figure 2-9. Team B--Moving in cross cover formation.

2-47. Fire Team A clears the hallway intersection, while Team B uses the Rolling-T technique to pass through Team A en route to the next room (Figure 2-10).

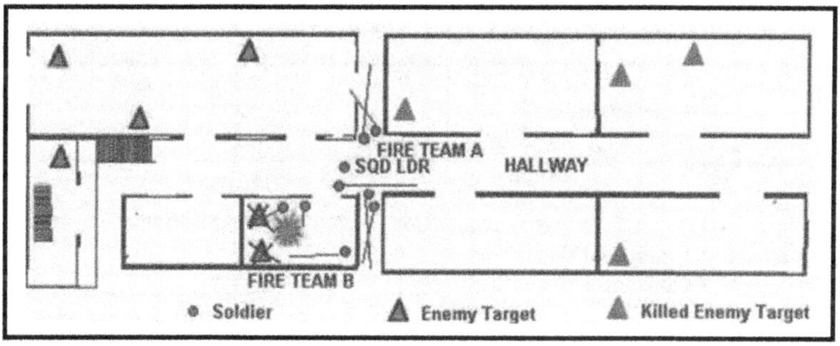

Figure 2-10. Team A--Clearing a hallway intersection.

2-48. Fire Team B throws the appropriate grenade, enters, engages the targets, and clears the room. Only three Soldiers enter, while the fourth Soldier provides rear security. Fire Team A prepares to move (Figure 2-11).

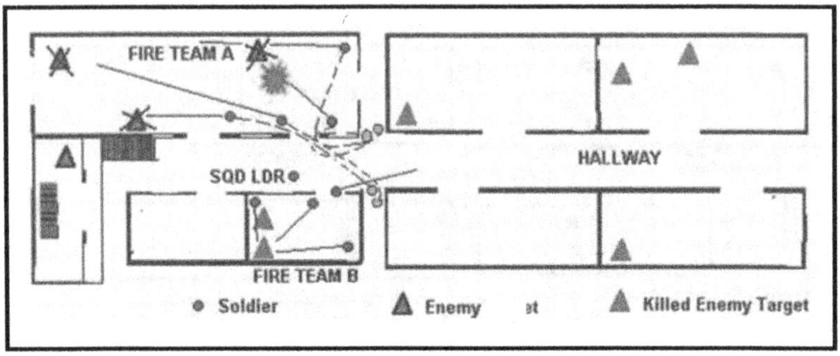

Figure 2-11. Team B--Clearing a room.

2-49. Fire Team A moves from the intersection and throws the appropriate grenade. The team enters, engages the targets, and then clears the next room. All team members enter, because of the size of the room. Fire Team B prepares to move (Figure 2-12).

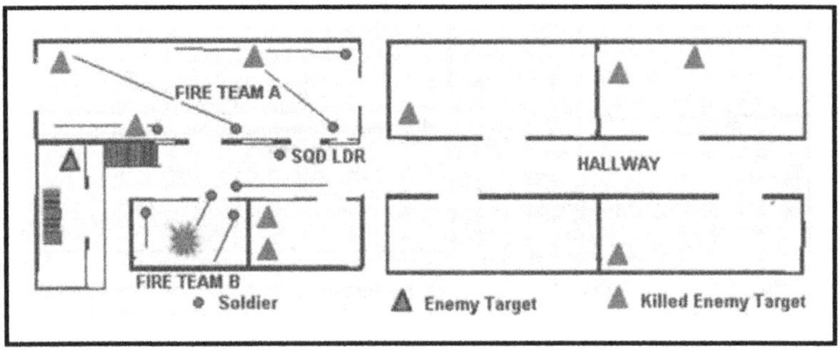

Figure 2-12. Team A--Clearing room by room.

2-50. Fire Team B employs the appropriate grenade, enters, and engages the target. Three Soldiers clear the next room. The fourth remains in the doorway and secures the rear. Fire Team A prepares to move up the stairs (Figure 2-13).

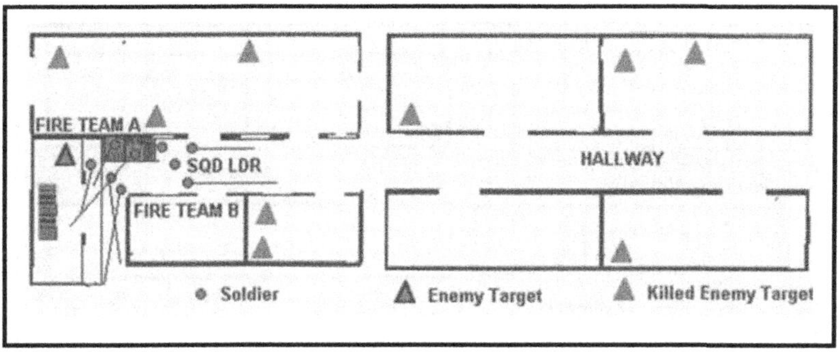

Figure 2-13. Team B--Clearing the next room.

2-51. Fire Team A moves up the stairs and onto the landing. The team prepares to enter the room off the landing. Fire Team B splits, and two members move around the corner and secure the area across from the stairs on the lower level. The other two members of Team B secure the hallway the squad came from (Figure 2-14).

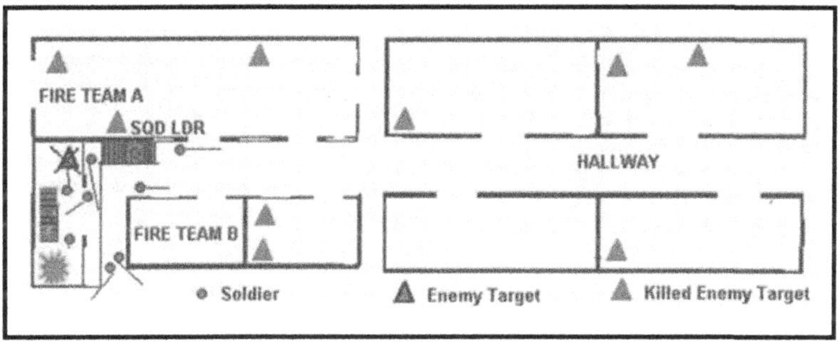

Figure 2-14. Securing the stairs and hallway.

2-52. Fire Team A members employ the appropriate grenade, enters the room, engages the target, and clears it. They find another set of stairs, and prepare to move up them. Fire Team B continues to secure the rear. The other team members move to the end of the short hallway and secure both directions (Figure 2-15).

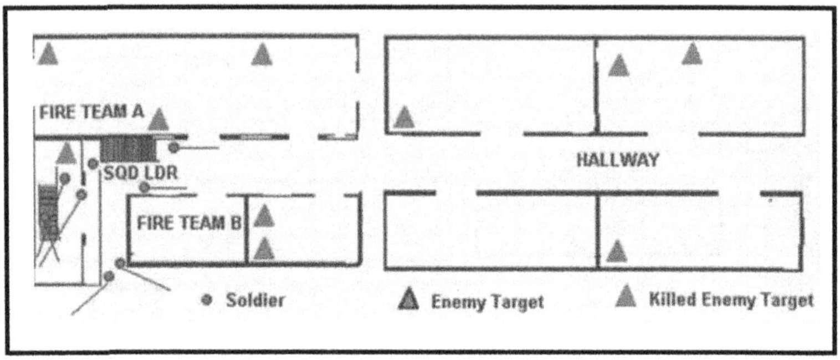

Figure 2-15. Preparing to clear more stairs.

2-53. Fire Team A clears the stairs and the landing. Fire team B continues to secure the ground level, ready to assist Team A, if necessary (Figure 2-16).

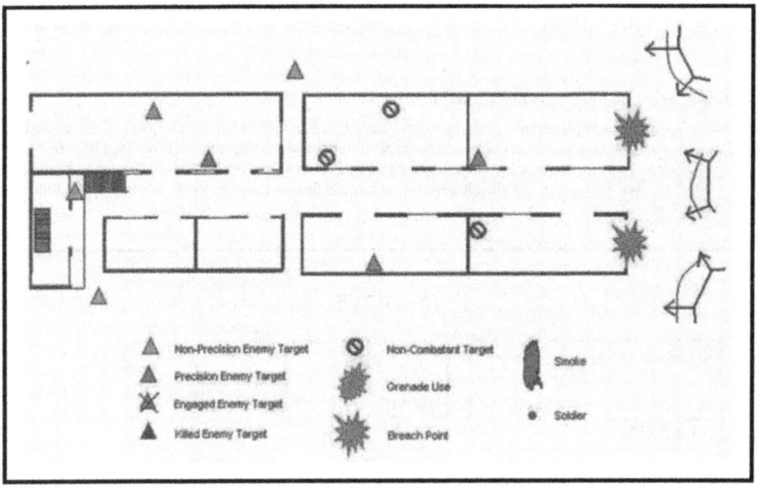

Figure 2-16. Team A--Clearing stairs and landing.

2-54. The previous example used no noncombatant targets, but used blanks/SESAMS and TES. This technique is the *crawl* phase for this station. Once the squad is ready to move on to more challenging exercises, the platoon leader/sergeant can add targets into the scenario. As proficiency increases, mixed combatant and noncombatant targets can be added. At any point, the platoon leader/sergeant can, until the platoon achieves proficiency, include multiple squads on the station. This increases the complexity of the scenarios.

Example Scenario--Preparing to Gain a Foothold

2-55. In this example platoon scenario, noncombatants are on the battlefield. The platoon has decided to clear through the buildings. They are going to try to do this without using the streets to move from building to building. Two squads will assault, and the other will remain in support/reserve. The support element isolates the buildings, while the assault team for each squad prepares and breaches each building in order to gain a foothold (Figure 2-17).

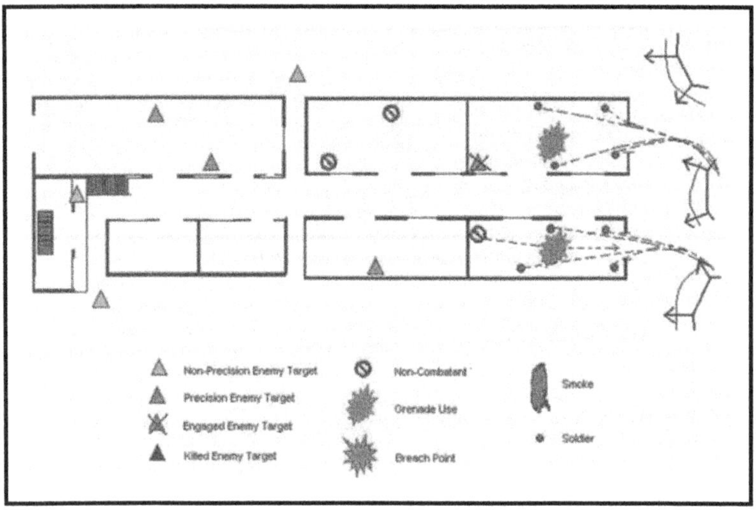

Figure 2-17. Preparing to gain a foothold.

2-56. The assault teams throw a grenade into the breach and their members follow, engaging enemy targets. They secure and move noncombatants to the collection point (CP) (Figure 2-18).

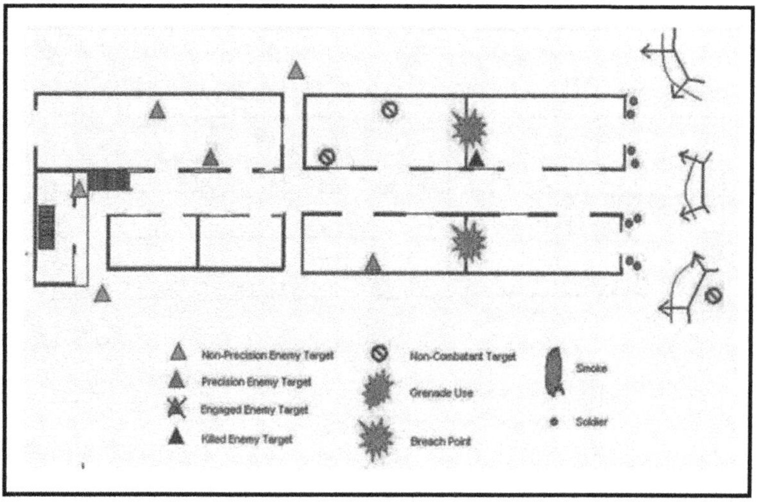

Figure 2-18. Securing and moving noncombatants to the command post.

2-57. The teams prepare and detonate wall breaches to continue movement through the building. Once the charges are ready for detonation, the assault teams move to a covered position and execute the breach (Figure 2-19).

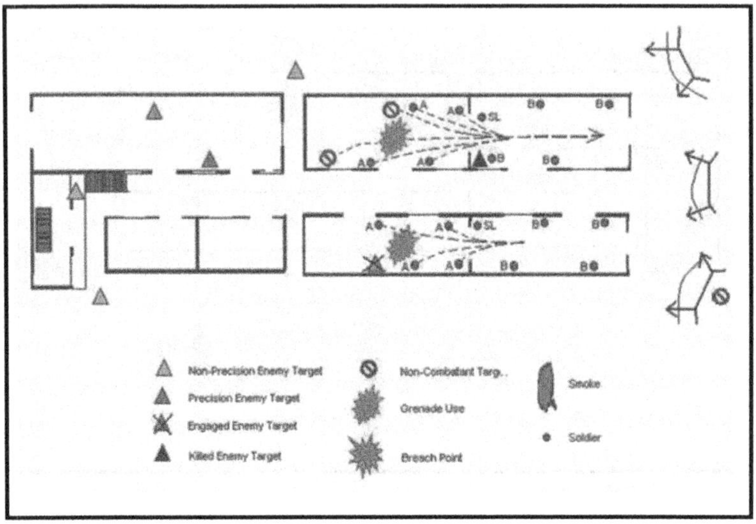

Figure 2-19. Executing breach.

2-58. The assault team employs grenades into the next room, enters the room, eliminates the enemy, and then secures and moves the noncombatants to a designated safe area. At the same time, the next assault team moves into the foothold (Figure 2-20).

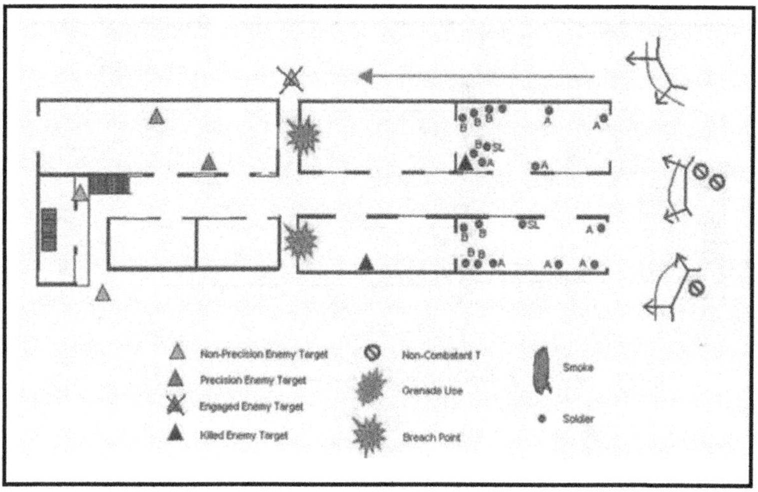

Figure 2-20. Moving to the foothold.

2-59. With the second room secured, another assault team moves forward and prepares to breach the wall Once the charges are placed, the teams move back to a covered position before detonating the charge. With the breach a success, the first nonprecision target is exposed and engaged by the support element (Figure 2-21).

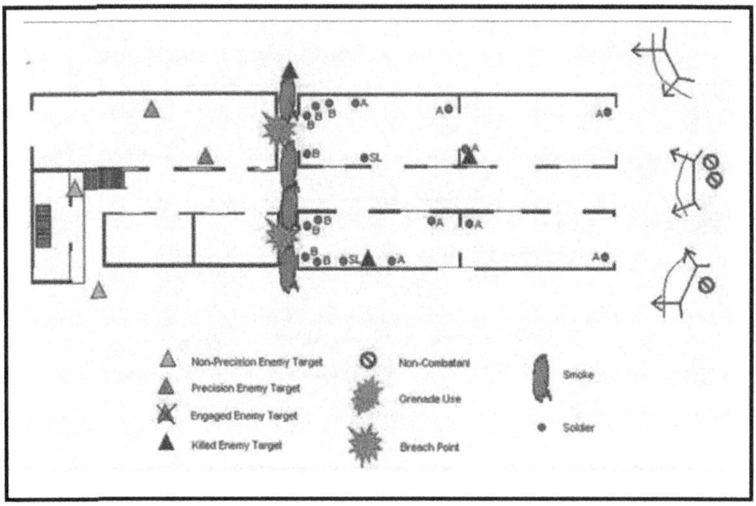

Figure 2-21. Breaching the wall.

2-60. Upon inspecting the breach, the assault teams find that they have an alley with another wall to breach. They employ smoke to obscure the teams placing the breaching charge. The first assault teams take up securing positions throughout the already cleared areas. Follow-on assault teams move to a covered position and execute the breach (Figure 2-22).

Note: Use smoke only in the alley, not inside the buildings.

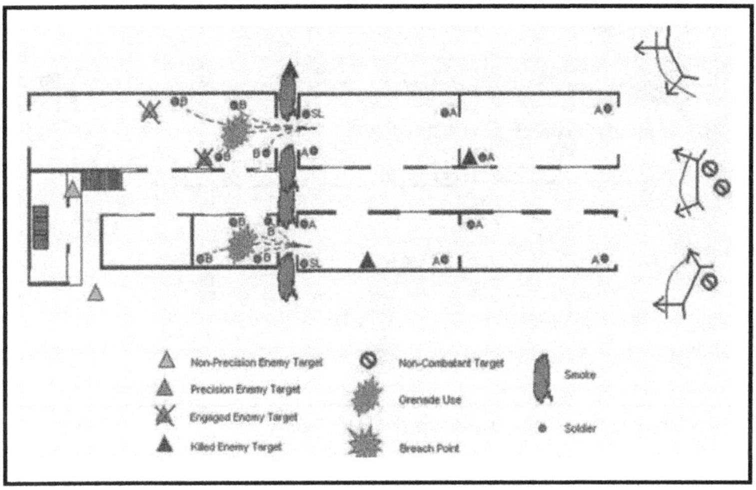

Figure 2-22. Securing a wall in an alley.

2-61. Once the assault teams execute the breach, they must determine if they need to use more smoke to cover their movement. Either way, once they have enough smoke to cover their movement, they throw in a grenade, enter, engage enemy targets, and clear the rooms (Figure 2-23).

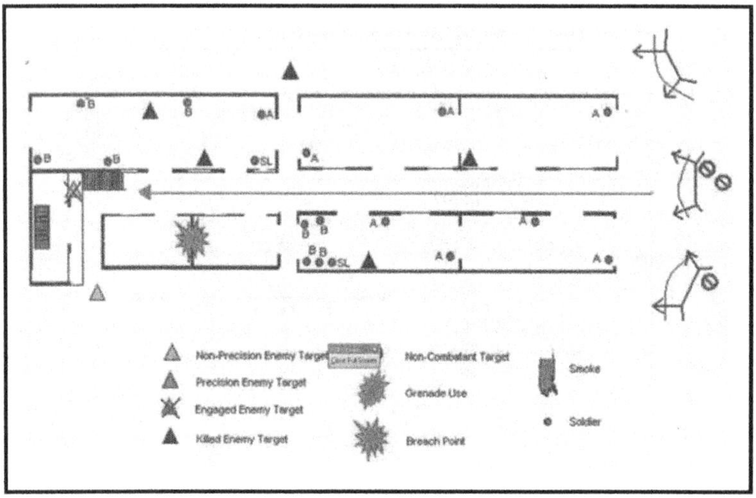

Figure 2-23. Obtaining obscurity.

2-62. While one team finds itself at the end of the building in their sector, the other team is faced with another wall and more clearing. One team prepares to breach the wall, and the other team takes up a hasty position and prepares for a counterattack. When the teams execute the breach, the next nonprecision target presents, and the support element engages it (Figure 2-24).

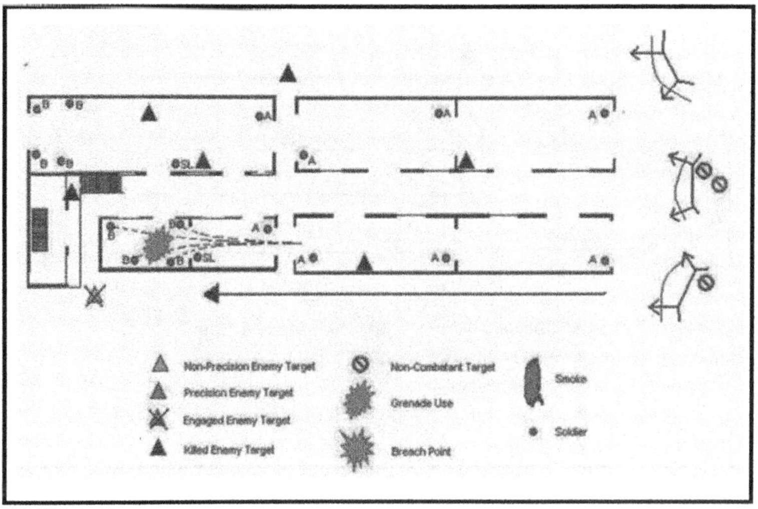

Figure 2-24. Clearing the wall.

2-63. Someone throws a grenade through the breach. The assault team enters and secures the room. When the grenade explodes inside the room, the final nonprecision target presents, and the support element engages it (Figure 2-25).

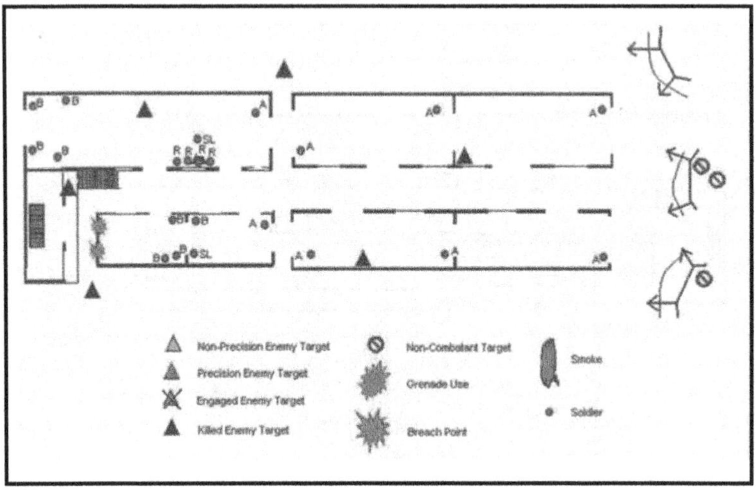

Figure 2-25. Securing a room.

2-64. With the first squad reaching its limit of advance, and with the second squad reaching the end of the building, the platoon leader calls another assault team forward from the reserve. The reserve assault team moves forward through the buildings, and positions itself to clear up the outside stairs of the two-story structure. The second squad places and detonates loophole charges on the far wall of the building. This supports the clearing of the last building (Figure 2-26).

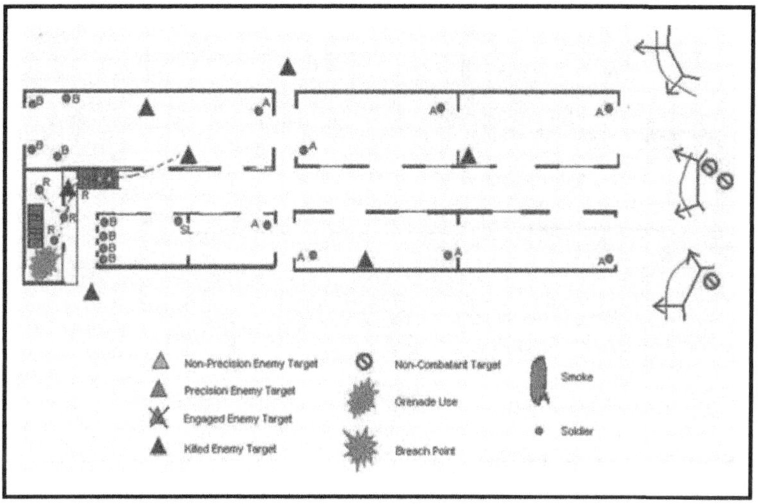

Figure 2-26. Reserve assault team moving forward.

2-65. Team B of the second squad moves into supporting fire positions. The new assault teams move up the outside stairs to the second floor of the final structure. They clear the balcony and the first room on the second floor. The support-by-fire position shift fires to the lower floor of the same structure (Figure 2-27).

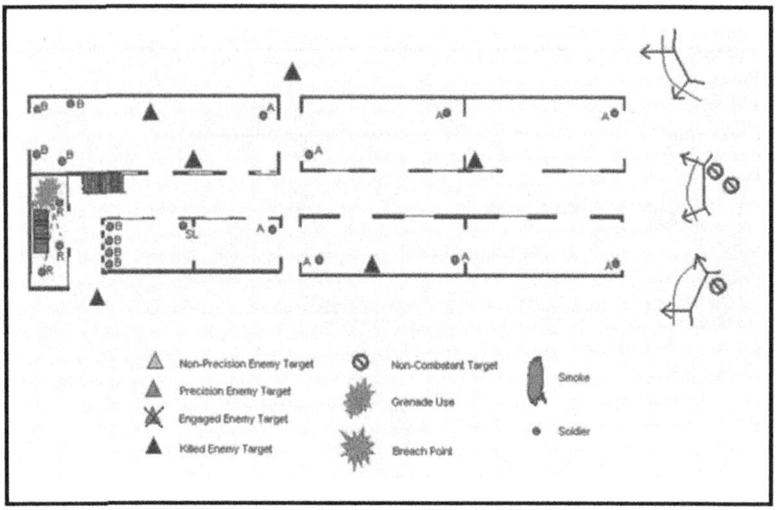

Figure 2-27. Team B, second squad--moving into the supporting fire position.

2-66. Once the floor is secure, the assault team moves down the stairs and clears the remaining floor (Figure 2-28). Meanwhile, the platoon repositions, and then consolidates and reorganizes.

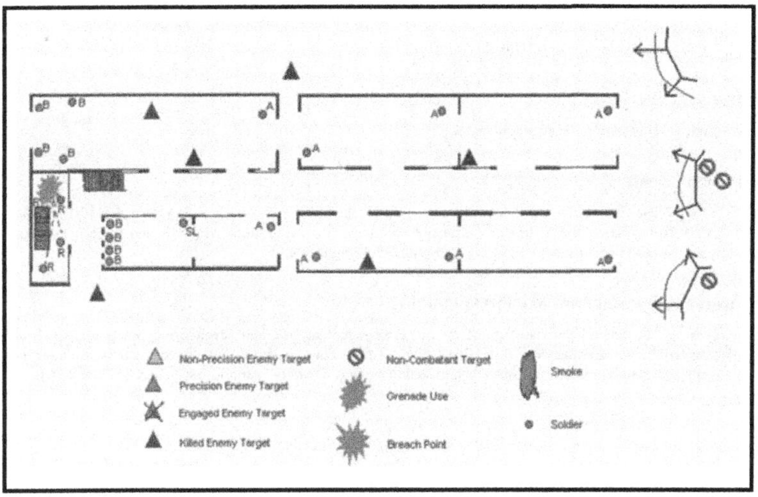

Figure 2-28. Consolidation and reorganization.

STATION 3: GRENADIER GUNNERY TRAINER

2-67. This is a live-fire station where M203 dual-purpose weapon gunners master the engagement of point-type M203 and M16 targets at 50 to 150 meters in an urban area. Tactical movement and fire commands are also trained at this station.

DESCRIPTION

2-68. This station has two main parts:

Wooden Facade

2-69. The first part is a wooden facade wall that looks like a two-story building. It is about 20 feet high by 30 feet wide. It has a doorway with outside steps, two lower floor windows and three upper floor windows, and two basement windows at ground level (Figure 2-29).

Target Engagement Lane and Firing Positions

2-70. The second part consists of the target engagement lane and seven firing positions, which are between the wooden façade and the first firing position. These positions have cover such as sandbags, log walls, and rubble. Firing points can be adjusted to support a variety of training scenarios (Figure 2-30).

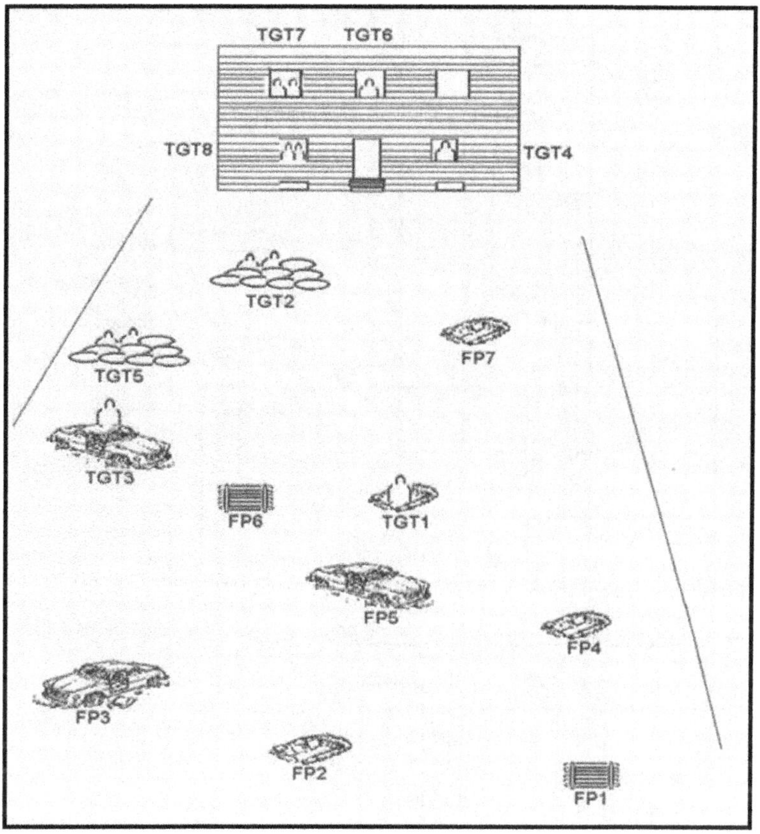

Figure 2-29. Part 1, Grenadier Gunnery Trainer, Station 3, wooden facade.

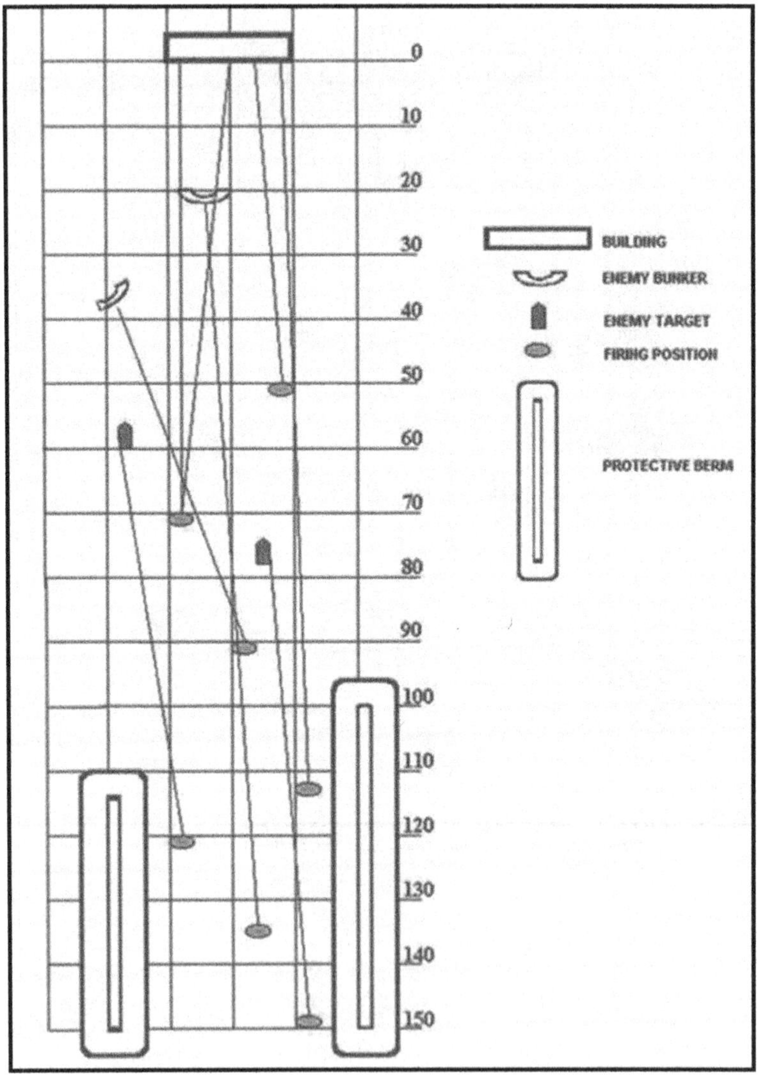

Figure 2-30. Part 2, Grenadier Gunnery Trainer, Station 3, target engagement lane and firing positions.

WARNING

Never engage targets with the 40-mm at ranges less than 31 meters, which is the minimum safe range. Trainers ensure the station accommodates this requirement as well as the 14-meter arming range.

TARGETRY

2-71. This station has twelve target mechanisms. These mechanisms are arranged so as to influence the gunner's decision as to which munition to use.

CONTROLLER REQUIREMENTS

2-72. As required by unit.

SAFETY REQUIREMENTS

2-73. Check the unit SOP and local safety regulations regarding live-fire ranges. Ensure all weapon systems are on SAFE before moving to the next firing position. Ensure that no 40-mm target is engaged at less than minimum arming range.

INDIVIDUAL AND COLLECTIVE TASKS TRAINED

2-74. These lists are noninclusive:

Supporting Individual Tasks

- 071-311-2007, *Engage Targets with M16-Series Rifle/M4 Carbine Rifle.*
- 071-311-2130, *Engage Targets with M203 Grenade Launcher.*
- 071-326-0503, *Move Over, Through, or Around Obstacles (except Minefields).*
- 071-326-0541, *Perform Movement Techniques during an Urban Operation.*
- 071-326-0557, *Select Hasty Firing Positions during an Urban Operation.*
- 071-326-0608, *Use Visual Signaling Techniques.*

Related Tactics, Techniques and Procedures

- FM 3-06.11, *Combined Arms Operations in Urban Terrain.*
  -- Chapter 3, *Offensive Operations.*
  -- Chapter 5, Fundamental Combat Skills.
- Appendix M, *Field Expedient Breaching of Common Urban Barriers.*

Training Support Requirements

2-75. Table 2-5 shows the minimum materials required to train at this station.

Table 2-5. Support requirements for Station 3.

Ammunition	*Rounds for each Weapon*	*Targets Required*
5.56-mm Ball	8	Four nonprecision human target mechanisms.
40-mm TP	8	Eight nonprecision human target mechanisms.

Conduct of Training

2-76. Whereas Table 2-5 showed the minimum amount of ammunition required for this station, Table 2-6 shows an example sequence of events and times, and Table 2-7 shows an example target scenario.

Table 2-6. Sequence of events and times for Station 3.

Event	*Action*	*Estimated Time*
1	*Perform Movement Techniques* *Select Hasty Firing Position* *Engage Targets With 5.56-mm (dry run)* *Engage targets with 40mm*	20 min
2	*Perform Movement Techniques* *Select Hasty Firing Position* *Engage Targets with 5.56mm (live-fire)* *Engage targets with 40mm*	10 min
NOTE: Training should include techniques outlined in FM 3-06.11		

Table 2-7. Example target scenario (presentation order) for Station 3.

Firing Position	*Weapon*	*Presentation Time*	*Range to Target*	*Target Presented*
1	5.56mm	6 sec	67 meters	Single Human Target (TGT 1)
2	40mm	2 min	105 meters	Multiple Human Targets (TGT 2)
3	5.56mm	6 sec	63 meters	Single Human Target (TGT 3)
4	5.56mm	6 sec	105 meters	Single Human Target (TGT 4)
5	40mm	2 min	57 meters	Multiple Human Targets (TGT 5)
6	40mm	2 min	70 meters	Multiple Human Targets (TGT 6)
6	5.56mm	6 sec	70 meters	Single Human Target (TGT 8)
7	40mm	2 min	50 meters	Multiple Human Targets (TGT 7)

STATION 4: URBAN OFFENSE/DEFENSE BUILDING, TASK/TECHNIQUE TRAINER

2-77. This station trains to attack or defend a building. Collective and individual TTP are trained. The platoon practices the task steps and performance measures for offensive and defensive tasks. This station increases the complexity of command, control, and maneuver (Figure 2-31).

Figure 2-31. Urban offense/defense building, Station 4.

DESCRIPTION

2-78. This station is a two-story building with a basement. Each floor has several rooms, windows, doorways, loopholes, and mouseholes. The gabled roof has windows and a hatch to the attic below and half of the rooftop is flat. The building has exterior and interior stairways.

TARGETRY

2-79. This station has ten target mechanisms.

CONTROLLER REQUIREMENTS

2-80. These are based on unit requirements.

SAFETY REQUIREMENTS

2-81. Only smoke, practice, and M84 stun grenades may be used at this station. It is not a live-fire station. Training is conducted using TES with blank 5.56-mm, or TES or SESAMS. Units should consult with their supporting combat engineers for the proper demolition effects simulators (DES) to use against the blow panels throughout the station.

DANGER

TO AVOID ASPHYXIATION, NEVER DETONATE SMOKE GRENADES INSIDE BUILDINGS. THEY ARE *ALL* DANGEROUS IN AN ENCLOSED SPACE.

WARNINGS

1. **To avoid injury, place your weapon on SAFE when climbing through obstacles such as windows, fences, and walls.**

2. **To avoid inflicting carbon burns or more serious injury, never fire blank ammunition at other Soldiers within 1 meter.**

DEMOLITION EFFECTS SIMULATORS

2-82. Demolition effects simulators (DES) replicate breaching to gain entry to buildings and rooms. The walls of the UAC, Shoot House, and CACTF facilities may be breached using wall breach charge DES. Generally, the interior and exterior walls in these facilities are made with blow panels, plasterboard, or similar material, and measure about 4 feet square.

2-83. The DES charges for use at the UAC, shoot house, and the CACTF are constructed to breach a predetermined entry point without damaging the facility.

Field-Expedient Charges

- Flex-linear shaped charge.
- Doorknob charge.
- E-silhouette charge.

Operational Charges

2-84. Unit trainers must consult with installation safety and combat engineers to determine the appropriate reduced charge. See also FM 3-06.11.

INDIVIDUAL OR COLLECTIVE TASKS TRAINED

Collective Task

- FM 3-21.8 (7-8).
  -- 07-3-9018, *Clear a Building (Platoon/Squad).*
  -- 07-6-1216, *Defend an Urban Area (Battalion/BCT).*

Related Collective Tasks

- 07-2-5027, *Conduct Consolidation and Reorganization (Company/Platoon).*
- 08-2-0003. *Treat Casualties.*
- 08-2-0004. *Evacuate Casualties.*
- 19-3-3100, *Process Detainees at a Collecting Point (CP).*
- 07-2-1342, *Conduct Tactical Movement (Company/Platoon).*
- 07-2-1189, *Conduct a Dismounted Tactical Road March (Company/Platoon).*
- 07-2-1198, *Conduct a Mounted Tactical Road March (Company/Platoon).*
- Warrior Battle Drill 12, *Enter and Clear a Room.*

Supporting Individual Tasks

- 052-193-1013, *Neutralize Booby Traps.*
- 071-010-0006, *Engage Targets with M249 Machine Gun.*
- 071-311-2007, *Engage Targets with M16-Series Rifle/M4 Carbine Rifle.*
- 071-315-2308, *Engage Targets with M16-Series Rifle/M4 Carbine Using a Night Vision Sight AN/PVS-4.*
- 071-325-4407, *Employ Hand Grenades.*
- 071-326-0501, *Move as a Member of a Fire Team.*
- 071-326-0503, *Move Over, Through, or Around Obstacles (except Minefields).*
- 071-326-0541, *Perform Movement Techniques during an Urban Operation.*
- 071-326-0557, *Select Hasty Firing Positions during an Urban Operation.*
- 071-326-0608, *Use Visual Signaling Techniques.*
- 071-326-5605, *Control Movement of a Fire Team.*
- 071-326-5611, *Conduct the Maneuver of a Squad.*
- 071-710-0008, *Operate Night Vision Goggles AN/PVS-7B.*
- 113-571-1022, *Perform Voice Communications.*
- 191-377-5250, *Handle Enemy Personnel and Equipment.*

Related Tactics, Techniques, and Procedures

- FM 3-06.11, *Combined Arms Operations in Urban Terrain.*
  -- Chapter 3, *Offensive Operations.*
  -- Chapter 5, *Fundamental Combat Skills.*
  -- Appendix M, *Field Expedient Breaching of Common Urban Barriers*
- FM 3-22.9, *Rifle Marksmanship.*

TRAINING SUPPORT REQUIREMENTS

2-85. Table 2-8 shows the minimum ammunition required to train at this station.

Table 2-8. Ammunition requirements for Station 4.

Ammunition	*Rounds for each Weapon*
5.56-mm Blank	16
5.55-mm Blank (Linked)	24
7.62-mm Blank	100
Practice Grenade Bodies	5
Practice Grenade Fuzes	10 per platoon
M84 Stun Grenade Bodies	5
M84 Stun Grenade Fuzes	10 per platoon
Smoke Grenades	4 per platoon

CONDUCT OF TRAINING

2-86. Table 2-9 shows the sequence of training for Station 4. Figures 2-32 and 2-33 show example scenarios of training events that can be conducted at this station. These are only examples. The unit should develop training events based on its METL and training needs.

Table 2-9. Sequence of events and times for Station 4.

Event	*Action*	*Estimated Time*
1	*Clear a Building*	2 Hours Crawl
		1 Hour Walk
		30 Minutes Run
2	*Defend Built-up Area/Building*	1 Hours Crawl
		1 Hour Walk
		2. Hour Run
3	*Conduct AAR*	1 Hour

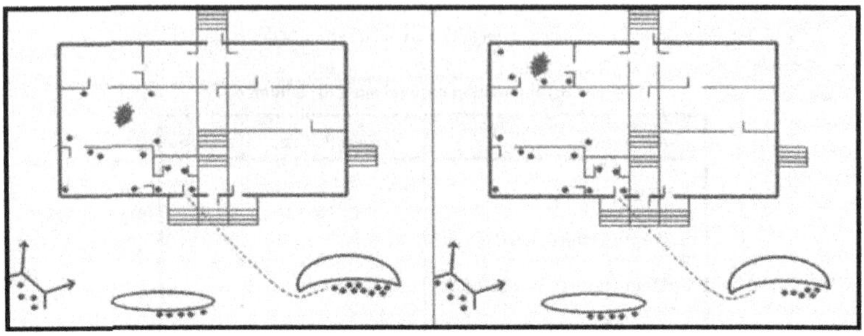

Figure 2-32. Event 1, *Clear a Building.*

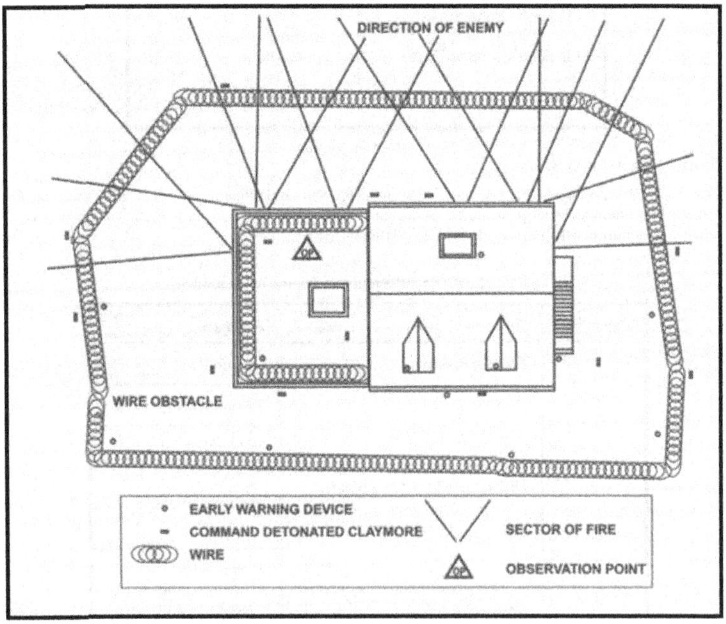

Figure 2-33. Event 2, *Defend a Building.*

STATION 5: UNDERGROUND TRAINER

2-87. This station provides training up to squad level for subterranean operations, specifically, the task *Clear and Move Tactically in a Subterranean Environment.*

DESCRIPTION

2-88. This station is a sewer system built in an irregular pattern, with four manhole covers. Two access tunnels lead from the outside tunnel to the underground loop, and the station is completely covered by dirt, only the manholes and ground level entrances are exposed (Figure 2-34).

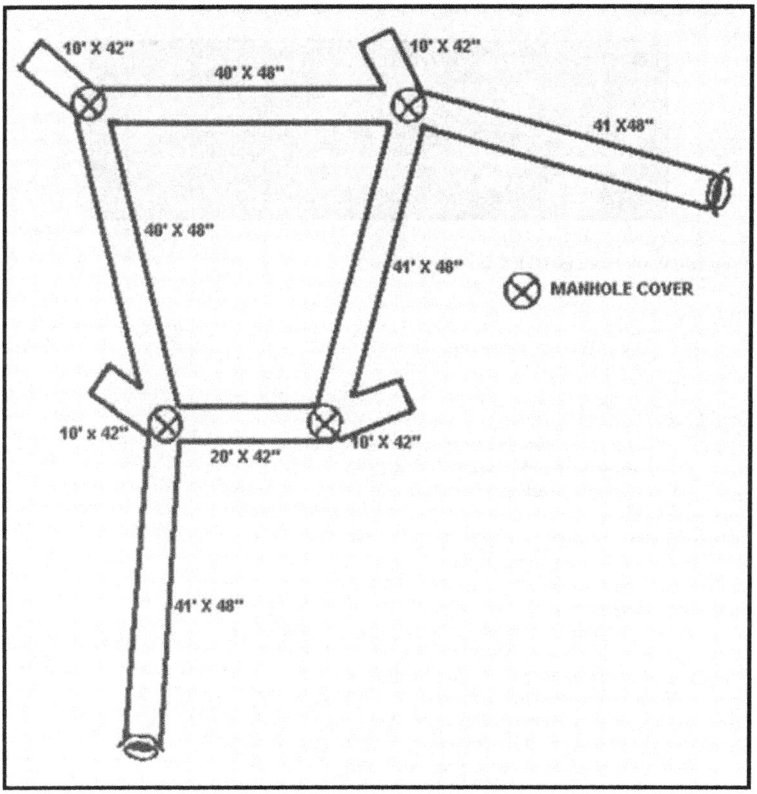

Figure 2-34. Underground Trainer, Station 5.

Targetry

2-89. Station 5 is a force-on-force training station and has no targetry.

Controller Requirements

2-90. As required by unit.

Safety Requirements

2-91. Soldiers should wear their helmets at all times when in the sewer system to avoid head injuries. Smoke grenades or M84 stun grenades *will not* be used in the underground trainer.

DANGER

TO AVOID ASPHYXIATION WHEN IN THE TRAINER, NEVER DETONATE SMOKE GRENADES IN THE SEWER. USING ANY TYPE OF EXPLOSIVE SIMULATOR IS DANGEROUS IN AN ENCLOSED SPACE. THE PROTECTIVE MASK IS NOT EFFECTIVE IN OXYGEN-DEFICIENT ATMOSPHERES.

Individual or Collective Tasks Trained

Collective Task

- *Move Tactically in a Subterranean Environment.*
- *Reconnoiter Area.*

Related Collective Tasks

- 07-2-1270, *Conduct an Infiltration (Company/Platoon).*
- 08-2-0003. *Treat Casualties.*
- 08-2-0004. *Evacuate Casualties.*

Supporting Individual Tasks

- 052-193-1013, *Neutralize Booby Traps.*
- 071-010-0006, *Engage Targets with M249 Machine Gun.*
- 071-311-2007, *Engage Targets with M16-Series Rifle/M4 Carbine Rifle.*
- 071-315-2308, *Engage Targets with M16-Series Rifle/M4 Carbine Using a Night Vision Sight AN/PVS-4.*
- 071-326-0501, *Move as a Member of a Fire Team.*
- 071-326-0503, *Move Over, Through, or Around Obstacles (except Minefields).*
- 071-326-0541, *Perform Movement Techniques during an Urban Operation.*
- 071-326-0557, *Select Hasty Firing Positions during an Urban Operation.*
- 071-326-0608, *Use Visual Signaling Techniques.*
- 071-326-5605, *Control Movement of a Fire Team.*
- 071-326-5611, *Conduct the Maneuver of a Squad.*
- 071-710-0008, *Operate Night Vision Goggles AN/PVS-7B.*

- 113-571-1022, *Perform Voice Communications*.
- 191-377-5250, *Handle Enemy Personnel and Equipment*.

Related Tactics, Techniques and Procedure

- FM 3-06.11, *Combined Arms Operations in Urban Terrain*.
  -- Chapter 3, *Offensive Operations*.
  -- Chapter 5, *Fundamental Combat Skills*.
  -- Appendix M, Field Expedient Breaching of Common Urban Barriers.
- FM 3-22.9, *Rifle Marksmanship*.

TRAINING SUPPORT REQUIREMENTS

2-92. Use Table 2-10 for the minimum ammunition required to train at this station.

Table 2-10. Ammunition requirements for Station 5.

Ammunition	*Rounds for each Weapon*
5.56-mm Blank	20
5.55-mm Blank (Linked)	40

CONDUCT OF TRAINING

2-93. Table 2-11 shows the sequence of training for this station. These are only examples. The unit should develop training events based on its METL and training needs.

Table 2-11. Sequence of events and times for Station 5.

Event	*Action*	*Estimated Time*
1	*Reconnoiter*	30 minutes (Crawl) 30 minutes (Walk) 30 minutes (Run)
2	*Conduct an Infiltration (Co/Plt)*	30minutes (Crawl) 30 minutes (Walk) 30 minutes (Run)
3	*Conduct AAR*	30 minutes

Chapter 3
Shoot House

This chapter describes the live-fire Shoot House and provides layouts and training scenarios. Fighting in built-up areas is fragmented; gains are often small. Therefore, units depend on the initiative, skill, and discipline of small-unit leaders and individual Soldiers. Leaders must be competent and confident in their urban operations skills and must have the courage to accomplish their missions while isolated from their parent units. Soldiers and leaders require mission-type orders that allow for decentralized execution.

PURPOSE

3-1. The Shoot House provides the leader with a facility to train and evaluate the unit during a live-fire exercise. Units are trained and evaluated on their ability to conduct tactical movement, engage targets, conduct breaches, and practice target discrimination in an urban environment.

DESCRIPTION

3-2. This facility has two parts, the live-fire facility and the AAR room:

LIVE-FIRE FACILITY

3-1. The live-fire facility is a single-level structure with a barn roof, catwalk, and eight adjoining rooms. Each room has video capability for AARs. The walls are bulletproof and prevent ricochets. Each side has an entrance. Replaceable blow panels throughout the facility can withstand reduced DES blasts. These also offer alternative entry points. A crane system in the roof allows trainers to selectively replace doorways and blow panels with solid wall sections. Before training explosive and mechanical breaching operations, trainers should place mobile exterior baffles or containment walls placed by each entry to prevent rounds and demolition effects from escaping. All of these capabilities let the unit vary its training scenarios. For safety, the Shoot House is windowless (Figure 3-1).

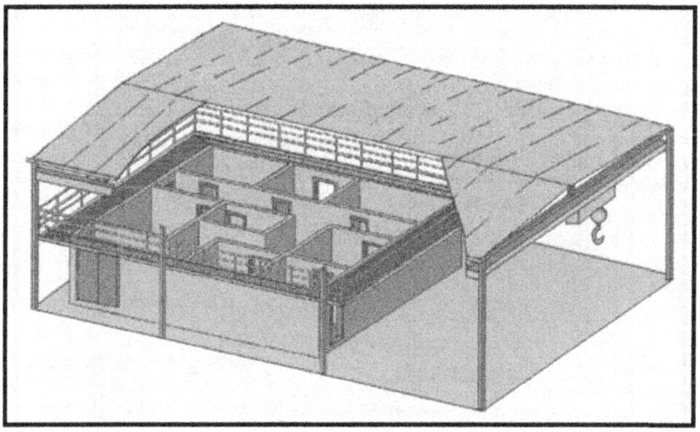

Figure 3-1. Shoot House.

AAR Building

3-2. The AAR building has a conference/AAR room that can accommodate up to a platoon size element, an instrumentation and targetry control room, and an AAR edit and projection room that can handle a formal AAR (Figure 3-2).

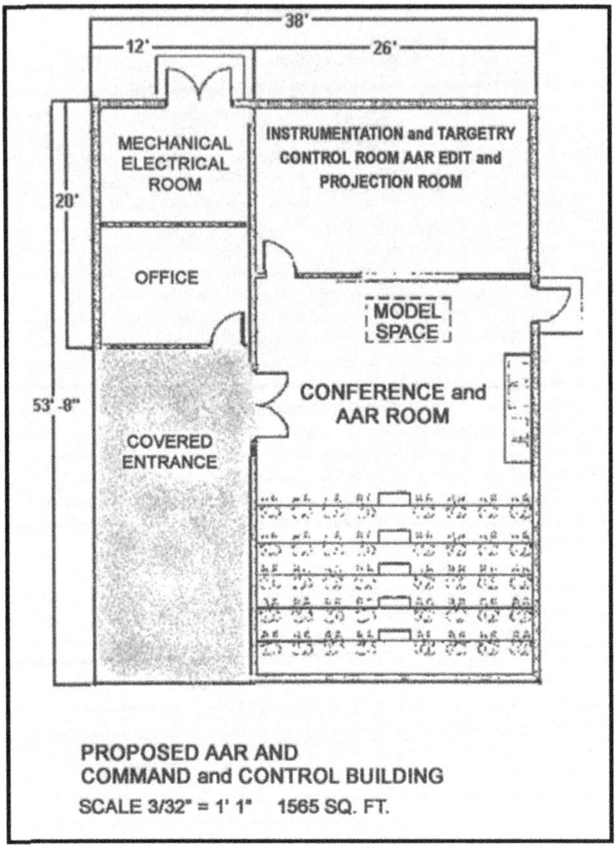

Figure 3-2. Command and control building and AAR room.

TRAINING AUDIENCE

3-3. The facility provides leaders the ability to train and evaluate the unit during a live-fire exercise. However, the unit's training requirements determine the size of the unit and the tasks to be trained at this facility. The unit develops its own organization and the steps for conducting the exercise based on its METL and training needs.

TRAINING INFORMATION

3-4. Tactical operations in an urban environment may range from combat missions to stability and support missions. Commanders should note that the Shoot House *is not* suited for the employment of live-fragmentation/concussion grenades, 40-mm HE/HEDP, antiarmor weapons, mortars, artillery, or air delivered munitions. Crew-served weapon systems, M2 and M1 tanks, and machine guns can be employed against supplemental targetry that is exterior to the Shoot House. The Shoot House has specially constructed areas that are used to conduct either mechanical or explosive breaching techniques. When conducting explosive breaching, units should consult with local Engineers for appropriate DES charges and consult Appendix A before conducting the training.

CONDUCT OF TRAINING

3-5. The unit develops realistic time frames for each operation or task. The times are based on the *crawl-walk-run* concept. The training event or STX can be arranged in any order as long as they follow a tactical scenario. *Assault a Building*, *Enter a Building and Clear a Room*, and *Clear a Building* are the primary collective tasks trained at this facility.

EXECUTION

3-6. Tables 3-1 through 3-5, pages 3-4 through 3-7, shows a sequence of events and times, and Figure 3-3 through Figure 3-6, pages 3-6 through 3-9, show example scenarios. These are only examples. The unit should develop training events based on its METL and training needs. However, the leader should do the following when conducting training at the live-fire Shoot House:

1. Identify the unit's critical missions (see commander's training guidance), and select those tasks that support specific tasks related to urban operations.
2. Arrange the related tasks in the order of their importance to mission accomplishment.
3. Assess the unit's current level of proficiency (strengths and weaknesses), and determine the degree of difficulty of each task. The training events and scenarios should start from the least difficult to the most difficult tasks. This gives the Soldiers confidence.
4. Identify the individual and leader tasks to be trained.
5. Conduct individual training first, because it is the first step in the building-block approach. It is also the basis for unit or collective task training. Mastery of individual skills is a precondition for effective training.
6. Set up training conditions based on unit proficiency. At first, keep the conditions basic, until Soldiers attain proficiency. Increase difficulty and realism as training progresses. Then, add limited visibility, CBRN factors, and live fires with service or short-range training ammunition (SRTA).
7. Understand that the Shoot House is a live-fire capable training facility. However, it can be used to train any level of proficiency, including blank, TES, SESAM, SRTA, and ball (no tracers). The leader assesses the unit's proficiency.

Table 3-1. Sequence of events and times for the platoon.

Element	*Event action*	*Estimated Time*
All 1st Squad 2d Squad 3rd Squad	1. Conduct precombat checks (for example; install, align, and troubleshoot MILES equipment; conduct inspection; draw equipment). 2. Support Operations. 3. Conduct Rehearsals. 4. Conduct Event 1 (Crawl).	2 Hours
3rd Squad 1st Squad 2d Squad	5. Support Operations. 6. Conduct Rehearsals. 7. Conduct Event 1 (Crawl).	2 Hours
2d Squad 3rd Squad 1st Squad	8. Support Operations. 9. Conduct Rehearsals. 10. Conduct Event 1 (Crawl).	2 Hours
1st Squad 2d Squad 3rd Squad	11. Support Operations. 12. Conduct Rehearsals. 13. Conduct Event 1 (Walk).	2 Hours
3rd Squad 1st Squad 2d Squad	14. Support Operations. 15. Conduct Rehearsals. 16. Conduct Event 1 (walk).	2 Hours
2d Squad 3rd Squad 1st Squad	17. Support Operations. 18. Conduct Rehearsals. 19. Conduct Event 1 (walk).	2 Hours
1st Squad 2d Squad 3rd Squad	20. Support Operations. 21. Conduct Rehearsals. 22. Conduct Event 1 (Run).	1 Hours
1st Squad 2d Squad 3rd Squad	23. Support Operations. 24. Conduct Rehearsals. 25. Conduct Event 1 (Run).	1 Hours
1st Squad 2d Squad	26. Support Operations. 27. Conduct Event 1 (Run).	1 Hours

Table 3-2. Event 1, Squad Sequence (*Crawl*).

Element	*Event Action*	*Estimated Time*
1 (CRAWL)	**Terminal Learning Objective** Familiarize the squad with how to enter and clear a room during the crawl phase of training **Tasks** *Enter and Clear a Room (Squad)* • *Perform Movement Technique* • *Select Hasty Firing Position* *Conduct a Breach (Mechanical)* • *Employ Hand Grenades* • *Engage Targets with M16A2/M4* • *Engage Targets with M249 Machine Gun*	**2 Hours**
NOTES: 1. Perform controlled pairs and target discrimination techniques (FM 3-22.9). **2. Include techniques (FM 3-06.11) and performance measures (FM 3-21.8 and DTMS).**		

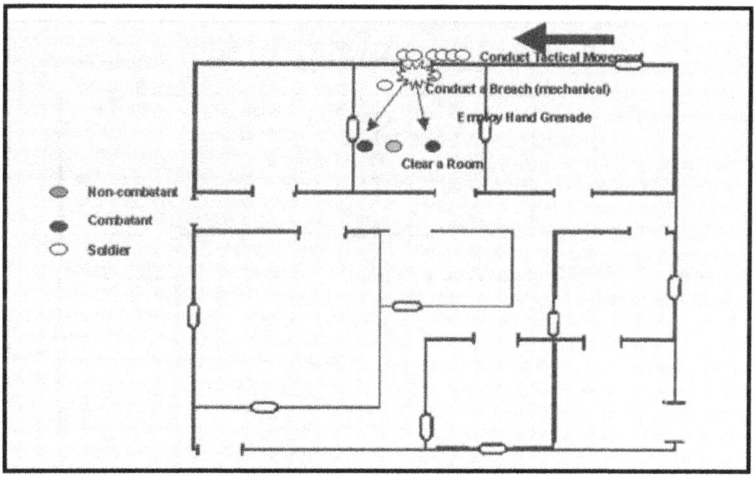

Figure 3-3. Event 1, *Squad Scenario (Crawl)*, force-on-targetry.

Table 3-3. Event 2, Squad Sequence (*Walk*).

Element	*Event Action*	*Estimated Time*
2 (WALK)	**Terminal Learning Objective** Familiarize the squad with how to enter and clear multiple rooms during the walk phase of training **Tasks** *Enter and Clear a Building (Squad)* • *Perform Movement Technique* • *Select Hasty Firing Position* *Conduct a Breach (Explosive DES)* • *Engage Targets with M16A2/M4* • *Engage Targets with M249 Machine Gun* *Conduct AAR*	2 Hours

NOTES: 1. Perform controlled pairs and target discrimination techniques (FM 3-22.9)
2. Include techniques (FM 3-06.11) and performance measures (FM 3-21.8 and DTMS).

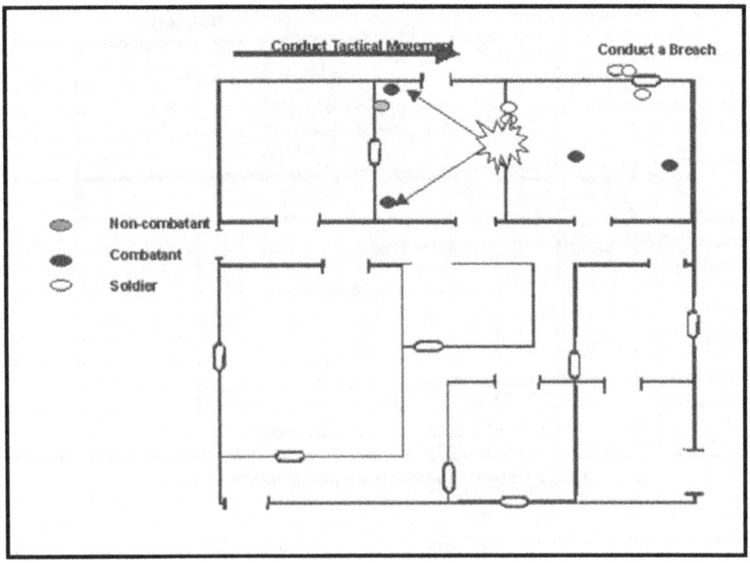

Figure 3-4. Event 2, *Squad Scenario (Walk)*, force-on-force.

Table 3-4. Event 3, Squad Sequence (*Run*).

Element	*Event Action*	*Estimated Time*
3 (RUN)	**Terminal Learning Objective** Refamiliarize squad with how to enter and clear multiple rooms during the run phase of training. **Tasks** *Enter and Clear a Building (Squad)* • *Perform Movement Technique* • *Select Hasty Firing Position* *Conduct a Breach (Mechanical)* • *Engage Targets with M16A2/M4* • *Engage Targets with M249 Machine Gun* *Clear a Hallway* • *Engage Targets with M16A2/M4* • *Engage Targets with M249 Machine Gun* *Conduct a Breach (explosive DES)* • *Engage Targets with M16A2/M4* • *Engage Targets with M249 Machine Gun* *Conduct AAR*	1 Hour

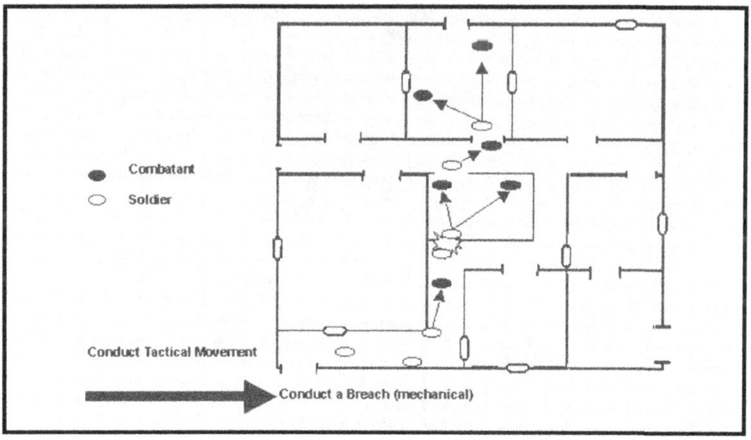

Figure 3-5. Event 3, *Squad Scenario (Run)*, force-on-targetry.

Table 3-5. Event 4, Platoon Evaluation Sequence (*Run*).

Element	*Event Action*	*Estimated Time*
4 (RUN)	*Occupy Assembly Area*	1 Hour
	Develop and Communicate a Plan	2 Hours
	Prepare for Combat	30 min
	Perform Passage of Line	1 Hour
	Conduct Tactical Movement	
	Break Contact	5 min
	Conduct Intermediate AAR	20 min
	Conduct Tactical Movement	
	Perform Overwatch/Support by Fire	1 Hours
	Assault a Building	1 Hours
	Consolidate and Reorganize	2 Hours
	Conduct Final AAR	1 Hours

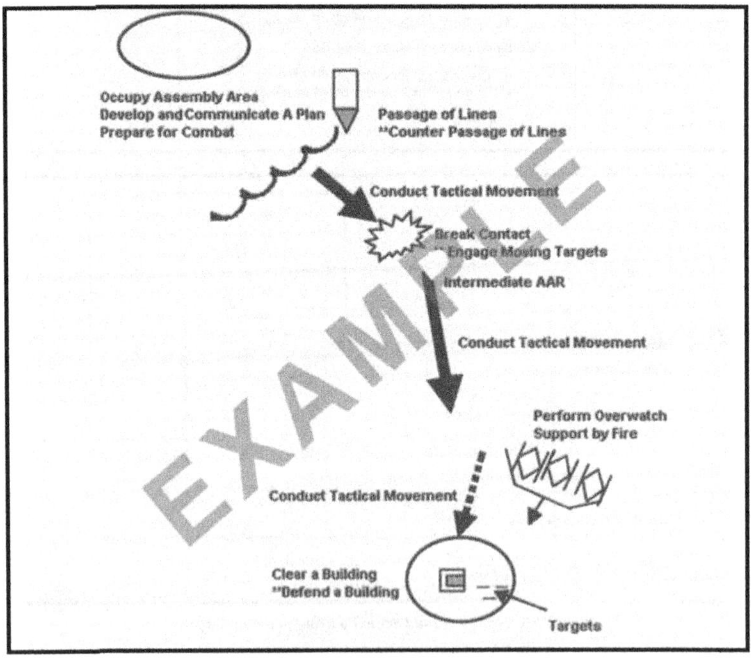

Figure 3-6. Example platoon tactical scenario (*Run*).

OPERATION ORDER

3-7. Figure 3-7 shows an example OPORD for the conduct of training.

```
                    ________________
                    (Classification)
                                          Copy ____ of ___ copies
                                        ____Company _____ Infantry
                                       (Unit Location)____________
                                        Date/time Group __________

(Message Reference Number)

OPORD: ______________
Reference Map, Series ____________________
Time Zone used Throughout the Order:______________________

TASK ORGANIZATION:

______Platoon Platoon
______Platoon ____ 60-mm Section
______Antiarmor Section
```

1. SITUATION

a. **Enemy Forces**. Annex B (Intelligence Overlay [prepared by battalion S-2]).

(1) The company is opposed by elements of the Red Company of the Blue Brigade at about 60 percent strength. Within our zone, we have a platoon-sized element occupying a defensive position in a building in the vicinity of (______). The enemy has small arms weapons and may call on company mortars for support.

(2) Current indications are the enemy is regrouping to initiate small-unit offensive actions in the area. His most likely course of action is to move to the (direction) to link up with elements of the parent unit.

b. **Friendly Forces**.

(1) (___) Infantry attacks at (______) hours to seize Objective ALPHA (grid). The intent is to prevent the enemy element from moving (direction) to link up with other enemy units.

(2) Company (__) on our (left, right) makes the main attack at (____) hours to seize Objective RED (grid) in order to establish blocking positions to stop the enemy's movement to the (direction).

____
(Classification)

Figure 3-7. Example company OPORD.

(Classification)

2. MISSION

Company (__) attacks to seizes Objective BLUE (grid) at (date/time) hours in order to prevent the enemy from regrouping and engaging Company (__) (main effort) from Objective BLUE.

3. EXECUTION

a. **Concept of Operation**. Annex C (Operations Overlay). (This should include those steps listed in the task , Conduct Troop-Leading Procedures, paragraph 17. These should be written out when possible and not just placed on an overlay.)

(1) Maneuver. (___) Platoon establishes a defense in its zone, NLT (date/time) hours, to destroy small enemy elements moving or exfiltrating to the (direction), from the vicinity of Objective BLUE. This defense supports the attack on Objective BLUE by completing destruction of any enemy that may escape. (Evaluated) Platoon conducts the company main attack at (hours) to seize Objective BLUE to destroy the enemy or force them to withdraw into (___) Platoon's defense. (___) Platoon supports the attack of the (Evaluated) Platoon on Objective BLUE to suppress enemy fires and isolate the objective. On seizure of Objective BLUE, (Evaluated) Platoon and (___) Platoon will defend in sector to destroy the enemy elements withdrawing to avoid (___) Company's attack.

(2) **Fire support**.

(a) Priority of fires.

- Initially to (Evaluated) Platoon.
- Upon seizure of Objective BLUE, priority shifts to (Defensive) Platoon.

(b) Priority targets.

- 60-mm priority target initially AA4031.
- On initiation of assault on Objective BLUE, priority shifts to target AA4027 to isolate the objective.
- Upon seizure of Objective BLUE, priority shifts to target AA4025.

(Classification)

Figure 3-7. Example company OPORD (continued).

(Classification)

b. **Tasks to maneuver units**.

(1) (___) Platoon. Destroy enemy elements moving to the (direction), NLT (date/time).

(2) (Evaluated) Platoon. Consolidate and defend in sector to destroy enemy elements following the seizure of Objective BLUE.

(3) (___) Platoon. Be prepared to assault Objective BLUE, on order, consolidate and defend in sector to destroy enemy elements following the seizure of Objective BLUE.

c. **Tasks to Subordinate Units**.

(1) 60-mm Section. Initially in the vicinity of CP 7, priority of fires to the (Evaluated) Platoon. Move with the (Supporting) Platoon. On order, priority of fires to the (Defensive) Platoon.

(2) Antiarmor Section. Attached to the (Supporting) Platoon.

d. **Coordinating Instructions**:

(1) (Defensive) Platoon reports defense established and locations.
(2) Platoon reports the seizure of all objectives.
(3) Platoon reports the consolidation and establishment of the defense.
(4) Order of march from the assembly area to the objective: defense platoon, evaluated platoon, headquarters, support platoon, 60-mm mortar.

4. **SERVICE SUPPORT**

a. **General**. Company trains collocated with the battalion combat trains in the vicinity of (grid). Supply distribution at the company CP.

(Classification)

Figure 3-7. Example company OPORD (continued).

(Classification)

b. **Materiel and Services**.

(1) Class I: Ration cycle MRE-MRE-MRE.

(2) Class V: Pick up basic load at company CP, NLT (date/time).

(3) Medical: Company casualty collection point is __________.

(4) EPW: EPW collection point is ___________.

5. **COMMAND AND SIGNAL**

a. **Command**. Company command group initially follows the (Evaluated) Platoon.

b. **Signal**.

(1) Radio-listening silence is in effect until contact is made.

(2) SOI _____ is in effect.

JONES
Commanding

AnnexES:
B-Intelligence Overlay
C-Operations Overlay
D-Fire Support Overlay

(Classification)

Figure 3-7. Example company OPORD (continued).

TRAINING SUPPORT REQUIREMENTS

3-8. Minimum trainers and evaluators should be based on unit proficiency, training needs, and local safety requirements. Table 3-6 contains consolidated support requirements.

Table 3-6. Consolidated support requirements.

Ammunition	*Event 1*	*Event 2*	*Event 3*	*Event 4 (Optional)*
5.56-mm (per weapon) ball, blank, SRTA, or SESAMS	20 Blank	20 SESAMS	20 Ball/SRTA	20 Ball/SRTA
5.56-mm linked (per weapon)	30 Blank	30 SESAMS	30 Ball/SRTA	30 Ball/SRTA
Hand grenade body (practice)	0	--	4	4
Hand grenade fuzes (practice) per platoon	0	--	4	4
Hand grenade, smoke, per platoon	0	6	6	6
M84 hand grenade, stun, per platoon	0	4	4	0
TES (MILES Equipment)				
M16 (1 per weapon, 1 per OPFOR)	NO	YES	NO	NO
M249 (1 per weapon)	NO	YES	NO	NO
SAAF (1 per platoon)	NO	YES	NO	NO
Controller Gun (1 per OC)	NO	YES	NO	NO
Targetry	5	0	5	10
OPFOR	NO	6	NO	NO

TRAINING AND EVALUATION

3-9. The T&EOs are the foundation of the MTP and the collective training of the platoon. T&EOs are training objectives (task, conditions, and standards) for the collective tasks that support platoon critical wartime operations. T&EOs may be trained separately, in an STX, in an FTX, or in live-fire exercises. For collective live-fire standards, the trainer needs to refer to the applicable Infantry Gunnery Manual for the appropriate course of fire. Those standards and courses of fire need to be integrated into the training exercise. Table 3-8 provides a list of related collective and supporting individual tasks.

Table 3-7. Related collective and supporting individual tasks.

Task	FM 3-21.8	FM 3-06.11	STP 21-1-SMCT	STP 7-11B1-SM-TB	STP 21-24-SMCT
Conduct Consolidation and Reorganization Treat Casualties Evacuate Casualties Process Enemy Prisoners of War/Captured Materiel Defend MOUT Building Conduct Tactical Movement	X	X			
Supporting Individual Tasks					
Engage Targets with an M16-Series Rifle			071-311-2007		
Engage Targets with an M16-Series Rifle Using a Night Vision Sight AN/PVS-4			071-315-2007		
Engage Targets with an M249 Machine Gun				071-010-0006	
Employ Hand Grenades			071-325-4407		
Move Over, Through, or Around Obstacles (Except Minefields)			071-326-0503		
Neutralize Booby Traps					052-193-1013
Move as a Member of a Fire Team				071-326-0501	
Perform Movement Techniques during MOUT		X		071-326-0541	
Conduct the Maneuver of a Squad				071-326-5611	
Conduct an Explosive Breach				071-440-0030	
Execute a Mechanical Breach				071-440-0027	
Conduct a Breach				071-440-0029	
Operate Night Vision Goggles AN/PVS-7				071-710-0008	
Select Hasty Firing Positions during MOUT				071-326-0557	
Control Movement of a Fire Team				071-326-5605	

Chapter 4
Breach Facility

Urban operations require Soldiers to enter buildings through walls, doors, windows, and roofs. The Breach Facility is designed to train Soldiers using tasks and techniques on how to breach locked doors, windows, and create man-sized holes in walls. The facility can be used to train mechanical, ballistic, thermal, and explosive breaching.

PURPOSE

4-1. The Breach Facility (Figure 4-1) is used to train Soldiers semiannually on the technical aspects of breaching techniques. It is also used to train TTP and explosive techniques not trained on any other type range.

Figure 4-1. Breach facility.

DESCRIPTION

4-2. This facility has three stations:

STATION 1

4-3. This station is a facade that is 8 feet high and 88 feet wide with four sections of two doors for a total of eight doors. Soldiers can conduct mechanical, thermal, ballistic, and explosive breaching techniques on doors at this station. Half of the doors open from the exterior, and the other half of the doors open from the interior (Figure 4-2).

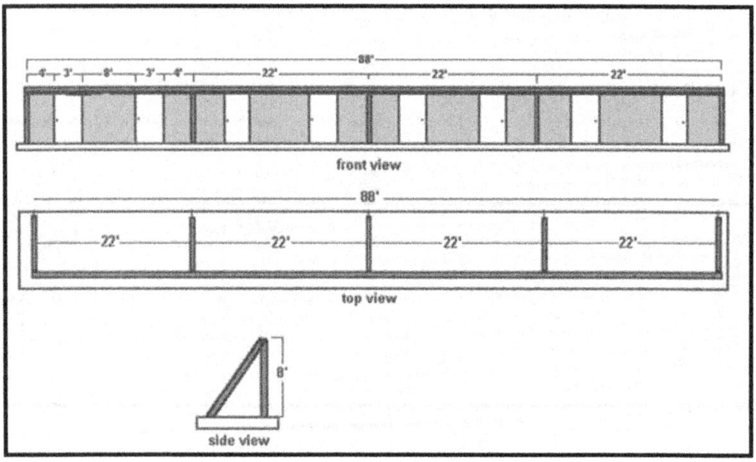

Figure 4-2. Station 1—Door breaching.

STATION 2

4-4. This station is a facade. It measures 10 feet high by 88 feet wide. It has four sections of two windows each, for a total of eight windows. Soldiers can conduct mechanical and explosive breaching techniques on windows at this station (Figure 4-3).

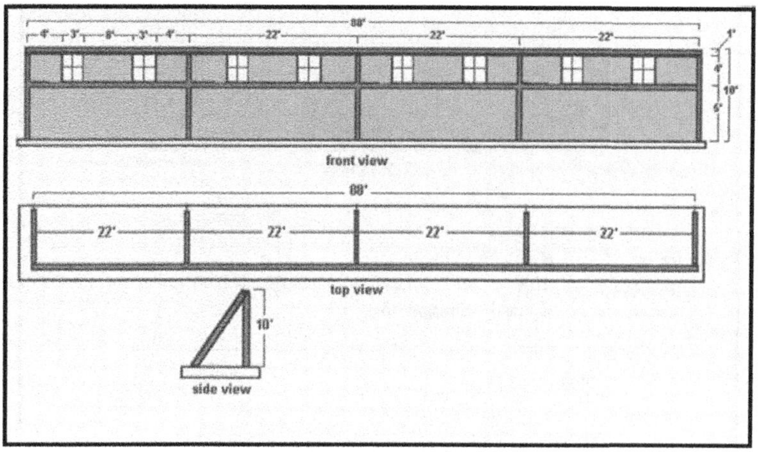

Figure 4-3. Station 2—Window breaching.

STATION 3

4-5. This station is a facade that looks like a wall of a building. It measures about 8 feet high by 28 feet wide. Precast panels will fit into each of the three sections. The panels measure 8 feet square and can be made of concrete, concrete blocks, stone, brick, wood or other material. The panels are designed for use with thermal and explosive breaching techniques (Figure 4-4).

TRAINING AUDIENCE

4-6. The facility provides Infantry, Engineer, MP, and Cavalry leaders with a facility to train and evaluate individual and leader breaching tasks. Unit training requirements determine the size of the unit and the tasks to be trained at this facility (Appendix B provides a unit training matrix). The individual tasks trained at the Breach Facility follow:

- 052-193-1003, *Prime Explosives Nonelectrically*.
- 071-440-0027, *Execute a Mechanical Breach*.
- 051-193-1055, *Construct a Nonelectric Initiating Assembly with Modernized Demolition Initiators (MDI)*.
- 052-193-1312, *Construct Demolition Initiating Sets*.
- *Conduct a Thermal Breach* (Engineers only).

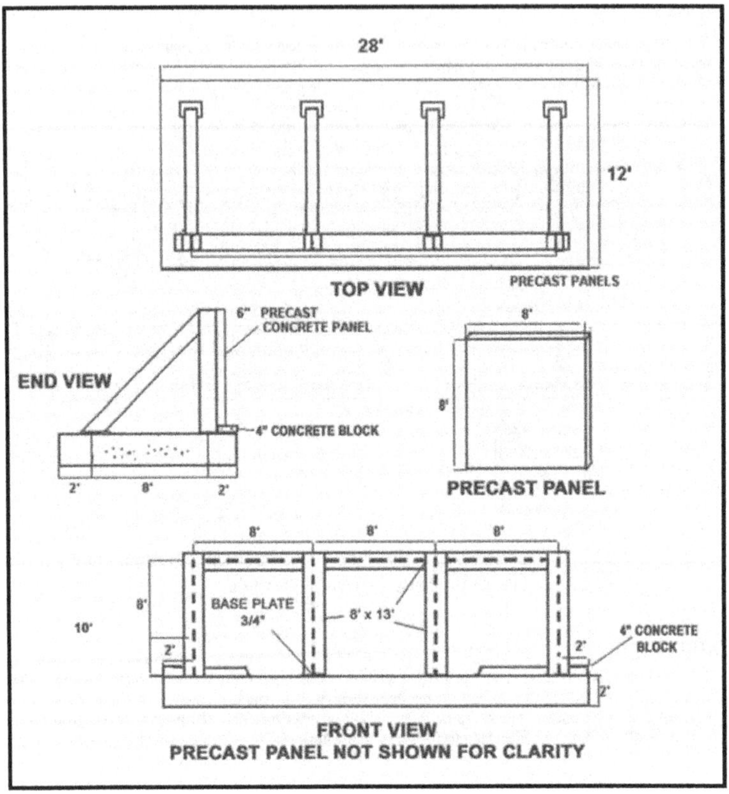

Figure 4-4. Station 3—Wall breaching.

TRAINING INFORMATION

4-7. Many types of organizations can train at this facility. Tactical operations in an urban environment may range from combat missions to stability and support actions. Commanders should note that this facility is an individual task and technique trainer. Training at the Breach Facility should be conducted prior to conducting collective training at the UAC, shoot house, and the CACTF. Trainers *must be certified* before conducting demolition training per local SOP or ACOM policy.

CONDUCT OF TRAINING

4-8. Trainers develop realistic time requirements for each task to be trained based on the *crawl-walk-run* concept and the size of the unit to be trained. The training event may be arranged in any order as long as it builds from simple to complex task and techniques. Table 4-1 contains an example sequence of events to assist units in developing their training plan. Leaders should do the following prior to conducting training. Before conducting ballistic breaching with automatic weapons, leaders should consult FM 3-06.11:

- Identify the unit's critical missions (see commander's training guidance) and select those tasks that support specific METL related to urban operations.
- Arrange the related tasks in the order of their importance to mission accomplishment.
- Assess the unit's current level of proficiency (strengths and weaknesses) and determine the degree of difficulty of each task. The training events should start from the less difficult tasks to the more difficult tasks. This instills confidence in the Soldiers conducting the training. Leaders should consult FM 3-06.11 for more information on explosive breach techniques.
- Identify the individual and leader tasks that are to be trained.
- Set up training conditions based on unit proficiency.

Table 4-1. Example platoon sequence of events.

Element	*Event Action*	*Estimated Time*
3d Sqd	Wall Breach: Explosive (crawl, walk, run)	3 hours
1st Sqd	Door Breach: Mechanical, ballistic, explosive (crawl, walk, run)	
2d Sqd	Window Breach: Mechanical, explosive (crawl, walk, run)	
2d Sqd	Wall Breach: Explosive (crawl, walk, run)	3 hours
3d Sqd	Door Breach: Mechanical, ballistic, explosive (crawl, walk, run)	
1st Sqd	Window Breach: Mechanical, explosive (crawl, walk, run)	
1st Sqd	Wall Breach: Explosive (crawl, walk, run)	3 hours
2d Sqd	Door Breach: Mechanical, ballistic, explosive (crawl, walk, run)	NOTE: Walls, doors, and windows require replacement after each run/iteration to train to standards.
3d Sqd	Window Breach: Mechanical, explosive (crawl, walk, run)	

TRAINING SUPPORT REQUIREMENTS

4-9. Table 4-2 shows minimum requirements based on three breach teams for each platoon-sized element. This facility is resource intensive, and Table 4-3 shows the total annual requirements per platoon-sized element.

Table 4-2. Consolidated minimum support requirements.

Ammunition/Equipment	*Crawl*	*Walk*	*Run*
Mechanical Breach: *Prying , Striking, or Cutting Tool*			
Door	(talk through)	1 per team (defeat one lock)	1 per team (defeat hinges)
Window	(talk through)	1 per team (defeat one lock)	1 per team (defeat hinges)
Ballistic Breach: Door			
Shotgun [1]	(talk through)	2 rds per weapon (defeat one lock)	5 rds per weapon (defeat hinges)
GREM (inert) [2]	(talk through)	1 per team – inert	1 per team - service
[1] If available to unit. [2] Grenade Rifle Entry Munition (GREM).			
Explosive Breach:			
Door: Det Cord (Flex Linear Charge) Nonelectric Firing System	(build)	(practice emplacement)	29 feet per team 1 per team
Wall: C4 (Satchel Charge) Nonelectric Firing System	(build)		4 blocks per team
		(practice emplacement)	23 feet per team 1 per team
Window: Det Cord (Flex Linear Charge) Nonelectric Firing System	(build)		29 feet per team 1 per team
		(practice emplacement)	
Thermal Breach Equipment Sappers (Only)	1	1	1

Table 4-3. Consolidated annual support requirements.

Ammunition/Equipment	*Total*
Doors	6
Windows	6
Wall Panels	6
Shotgun Rounds	42
GREM	6 inert/2 service
Det Cord (feet)	486
C4 (blocks)	24
Nonelectric Firing Systems	18

TRAINING AND EVALUATION

4-10. Individual training is the first step in the building-block approach and sets the basis for unit or collective training. Mastery of individual skills is an essential precondition for urban operations training and a good method for conducting individual training is through concurrent training. Concurrent training is designed to train groups of Soldiers on different tasks or techniques and these tasks or techniques may or may not be related. For example, a leader may subdivide the unit at the Breach Facility into breaching orders. One group of Soldiers may be conducting an explosive breach, while other groups are conducting mechanical and ballistic breaches. Ballistic breaching and explosive breaching does have a SDZ, an MSD, and ricochet considerations.

SAFETY CONSIDERATIONS

4-11. Safety requires effort on the part of everyone. It is not enough to teach safety rules; the rules must be practiced and enforced. Depending on the nature of the obstacle, the breacher may use mechanical, ballistic or explosive techniques. The following guidelines should be followed when training at the Breach Facility.

- Wear all appropriate personal protective equipment and follow appropriate guidelines:
  - -- Safety glasses.
  - -- Hearing protection.
  - -- Body armor.
  - -- Helmet.
  - -- Gloves.
  - -- Elbow and knee pads.
- Inspect equipment before use.
- Use appropriate tool.
- Use appropriate amount of explosive not to exceed four blocks of C4. Place the C4 on the wall only, not on the foundation, as this would cause irrepairable damage.
- Stay within the surface danger zone.
- Always use the minimum safe distance.
- Do not modify tool.
- Ensure proper use and maintenance of hydraulic and pneumatic tools; tools can burst under pressure when improperly used.

Chapter 5

Combined Arms Collective Training Facility

The home station CACTF is designed to conduct multiechelon, full-spectrum operations training up to battalion TF level. Example company and platoon STX lanes, and stability and support operations tasks, are provided to help units develop training plans. Leaders use training exercises to train, evaluate, and practice performance-oriented collective tasks. TEWT, STX, and FTX are the three types of exercises covered in this chapter. Each is designed to help the trainer develop urban operations training skills and increase unit proficiency in an urban environment. The urban operations collective training exercises were developed using the standard CACTF as described in this chapter and may be modified to fit available urban operations facilities.

PURPOSE

5-1. The leader or commander is provided with a CACTF to evaluate unit urban operations proficiency. The CACTF is intended to support blank fire, MILES/TES, SESAMS, STX, and FTX scenarios on a semiannual basis. The CACTF supports branch-specific lane training and combined arms training up to battalion level across the full spectrum of operations: offense, defense, stability, and support.

DESCRIPTION

5-2. The CACTF replicates an urban environment. The facility consists of 2.25 square kilometers of urban sprawl with 20 to 26 buildings, roads, alleys, parking areas, underground sewers, parks, athletic fields, and command and control building. The actual size and configuration of the CACTF depends on the local installation site requirements. The CACTF is designed to support heavy and light Infantry, Armor, artillery, and aviation positioning and maneuver. Table 5-1 shows the recommended types of buildings and features in the CACTF.

Table 5-1. CACTF characteristics.

20 to 26 Buildings (1.5 km by 1.5 km)	
Tunnel and sewer system	Props and furniture
Shantytown	Targetry
One 3-story building	Audio and image capture with EDIT/REPLAY capability
Three 2-story buildings	Control building with AAR facility
Industrial area	Breachable walls
Electricity and potable water	
City dump	

5-3. The buildings have one, two, or three stories (some with basements), sloped and flat roofs. CACTF building variations pose different tactical and technical training challenges. Figure 5-1 shows an example CACTF containing 24 buildings with the following characteristics:

- The hotel is a dominating structure typical of a central business district. The hotel has an elevator shaft, fire escapes, and a large first-floor lobby.
- The two-story office building contains a series of identical rooms.
- The school has a long, central corridor, large windows, and a single large room.
- The townhouse has multiple sections with a common attic.
- The service station accommodates combat vehicles in its service bay.

5-4. The buildings have other features that enhance the CACTF's training value.

- Constructed mouse-holes permit movement between rooms and floors.
- Loopholes in roofs and outside walls allow observation and fields of fire.
- Roof hatchways lead to the top floors of multiple story buildings.

5-5. Rooms vary in size, and in number of doorways and windows. These variations require Soldiers to identify rooms that provide suitable indoor fighting positions for TOW, Javelin, Dragon, and AT4 antitank weapon systems.

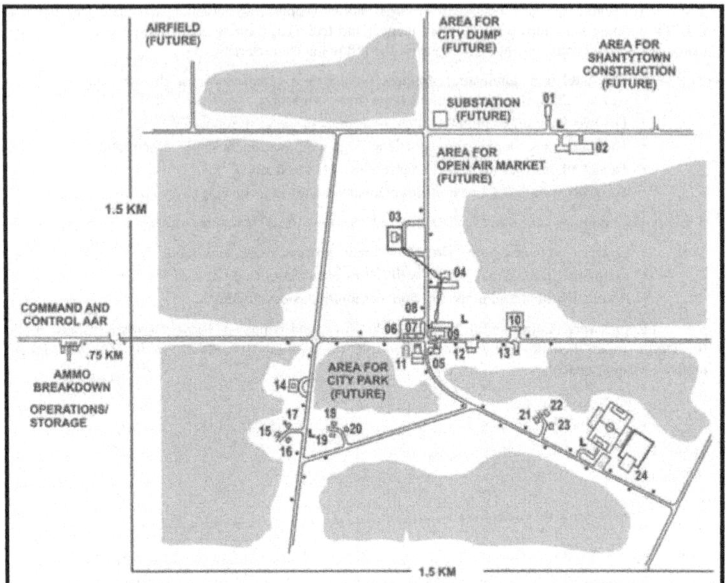

BLDG #	SIZE AND TYPE
01	1 STORY WAREHOUSE
02	1 STORY WAREHOUSE
03	3 STORY OFFICE (EMBASSY)
04	2 STORY OFFICE
05	1 STORY SERVICE STATION
06	1 STORY BUSINESS
07	1 STORY BUSINESS
08	1 STORY BUSINESS
09	3 STORY HOTEL
10	1 STORY POLICE STATION/JAIL
11	CHURCH AND CEMETARY
12	1 STORY BUSINESS
13	1 STORY BANK
14	2 STORY TOWNHOUSE
15	1 STORY RESIDENT
16	1 STORY RESIDENT
17	1 STORY RESIDENT
18	1 STORY RESIDENT
19	1 STORY RESIDENT
20	1 STORY RESIDENT
21	1 STORY RESIDENT
22	1 STORY RESIDENT
23	1 STORY RESIDENT
24	2 STORY SCHOOL

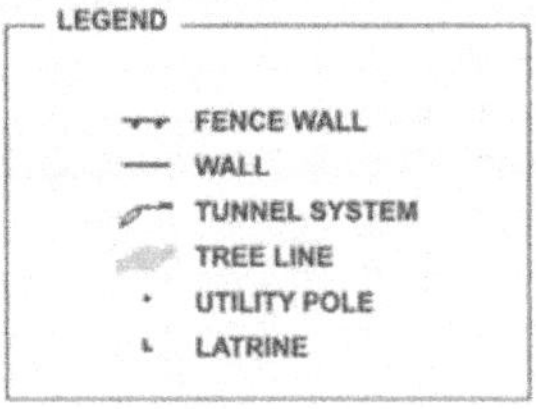

SCALE: 1" = 500'

Figure 5-1. Example CACTF schematic.

5-6. Buildings are arranged in a realistic urban pattern. The CACTF represents an area consisting mainly of residential, commercial, public institutional, and light industrial buildings. Buildings and streets are arrayed with sufficient dispersion to allow up to nine platoon-sized STX sectors (Figure 5-2) or six company/team STX sectors (Figure 5-3).

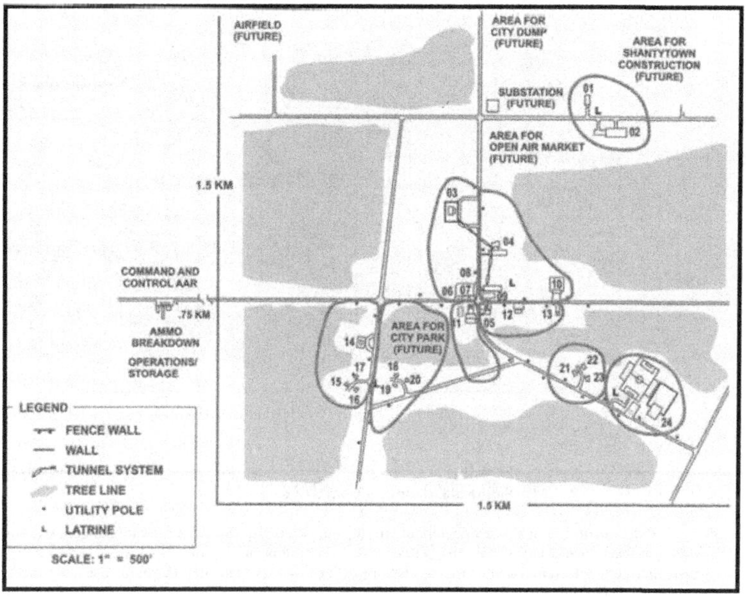

Figure 5-2. Platoon ODSS STX sectors.

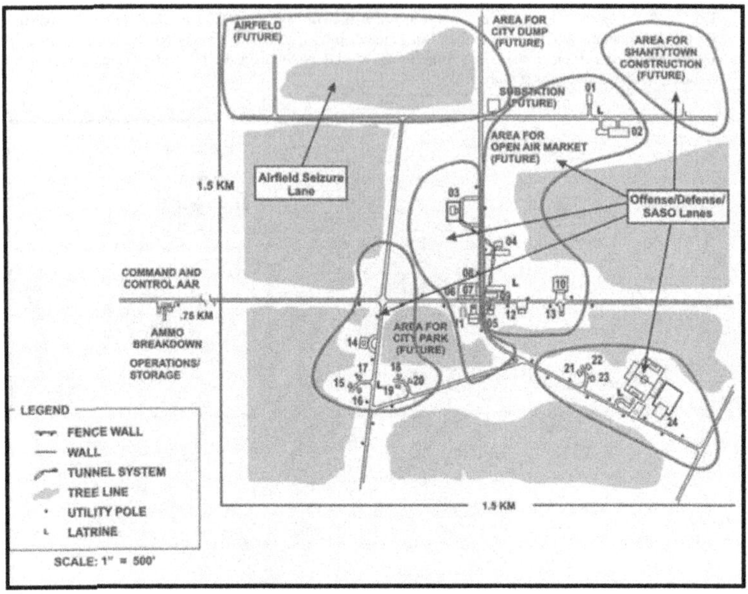

Figure 5-3. Company/team STX sectors.

5-7. Other urban features pose a variety of terrain considerations. The street network includes one and two-lane primary, secondary, local, and service roads. They join in T-shaped, L-shaped, and four-way intersections. Other features may be used to enhance realism such as signs, benches, furniture, newsstands, utility poles, street lights, vegetation, simulated rubble, live civilians on the battlefield, role players, and OPFOR. Open areas next to the CACTF can be used for maneuvering and staging areas. The specific details and layout of the CACTF may vary from one installation to another.

CAUTION

Trainers must ensure that fighting position reinforcement does not exceed the weight limits of the buildings in order to avoid structural damage and safety hazards to the Soldiers.

TACTICAL EXERCISE WITHOUT TROOPS

5-8. A terrain walk and a TEWT may be conducted in a local downtown area to prepare leaders and to take advantage of the available time in the CACTF. Terrain walks and TEWTs are conducted to teach leaders the tactical value of urban terrain. Appendix A provides training tips and additional information regarding the conduct of a TEWT.

COLLECTIVE TRAINING EXERCISES

5-9. The following exercises were developed IAW the guidance in FM 7-0, FM 7-1, FM 25-4, and TC 7-98-1. These publications include offensive and defensive scenarios for combat in urban areas, specifically, two platoon STXs and two company FTXs. Each can be conducted with blank ammunition; with MILES/TES; with 9-mm or 5.56-mm SESAMS for short-range precision target engagements (less than 50 meters); or force on force. Each was chosen to help the commander develop, sustain, and evaluate his unit's mission proficiency:

- Infantry Rifle Platoon Attack STX
- Mechanized Infantry Platoon Defend STX
- Infantry Rifle Company Attack FTX
- Mechanized Infantry Company Team Defend FTX

INFANTRY RIFLE PLATOON STX ATTACK

1. **Objective.**

This example STX trains collective, leader, and individual tasks for the platoon operation, Attack in a Built-Up Area.

2. **Interface.**

This STX is supported by two warrior drills, *React to Contact* and *Enter and Clear a Building*.

3. **Training.**

a. *Guidance*. FM 7-0 and FM 7-1 provide training guidance, tips, and enhancers for planning and executing STXs. The trainer should review the individual, leader, and collective tasks to be performed during the STX to determine the ones that require initial or refresher training.

b. *General Situation*.

(1) The exercise scenario follows: Contact with the enemy has been reestablished. Initial reports indicate he is at 65 to 70 percent strength, and that he has not received reinforcements. His defensive positions are in the CACTF and are not well established. He has indirect fire and CAS; he has already used chemicals, and will probably do so again. The platoon receives an order to prevent the enemy from establishing a heavily fortified defense in the CACTF. Such a defense could prevent friendly forces from continuing offensive operations.

(2) This exercise begins when the platoon leader receives the company FRAGO, and it ends after the platoon consolidates and reorganizes on the objective. An AAR should be held after completing the task, *React to Contact*, and at the completion of the exercise. If necessary, portions of the exercise should be repeated until the platoon performs them to standard. Figure 5-4 provides a graphic scenario of task performance for the STX *Attack*. Table 5-2 provides a recommended sequence for performance of T&EOs, and the time allotted for each portion of the STX.

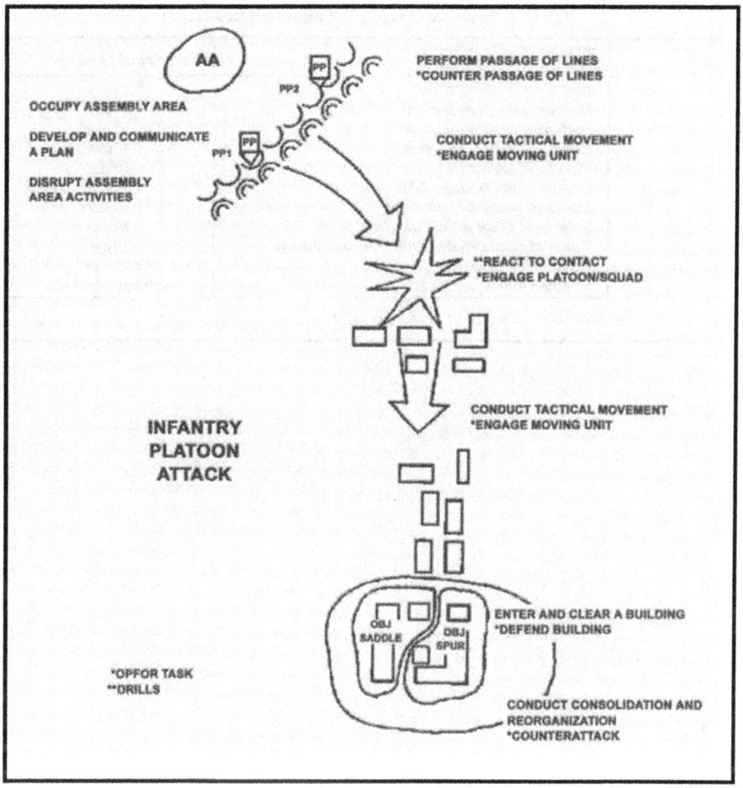

Figure 5-4. Attack STX scenario.

Table 5-2. Sequence of events and times for STX Attack.

Event	Action	Estimated Time
1	*Occupy Assembly Area*	1 hour
2	*Develop and Communicate a Plan*	1 hour
3	*Perform Passage of Lines*	1 hour
4	*Conduct Tactical Movement*	1 hour
5	*React to Contact (Drill)*	30 min
6	*Conduct Intermediate AAR*	30 min
7	*Conduct Tactical Movement*	30 min
8	*Enter and Clear a Building (Drill)*	1 hour
9	*Conduct Consolidation and Reorganization*	1 hour
10	*Conduct Final AAR*	1 hour
	* **Total Time**	**8 hours, 30 min**
* Conducting an exercise at night or in increased MOPP requires additional time.		

c. *Special Situation.* The platoon is part of a company in a secure assembly area when a FRAGO to attack is received (Figure 5-5).

1. **SITUATION.** The company is opposed by one platoon.

a. *Enemy Forces.* The enemy is at 65 to 70 percent strength. He is preparing a defense in CACTF (**MOUT site**) to prevent friendly forces from continuing to attack along the CACTF avenue of approach. He is expected to use artillery-delivered nonpersistent nerve agents in the defense of CACTF.

b. *Friendly Forces.* (Battalion designation) attacks (**date-time group**) to destroy enemy forces at OBJ COWBOY (**CACTF**) in order to pass follow-on forces, allowing friendly forces to continue the attack.

2. **MISSION.** (___) Company seizes OBJ HORSE (CACTF) vicinity (**grid**) NLT (**date-time group**) to prevent the enemy from establishing a heavily fortified defense.

3. **EXECUTION.**

a. *Concept of the Operation.* (See overlay.). The company moves in a wedge to Assault Position 1, assault CACTF to seize OBJ HORSE, consolidate, and reorganize.

(1) Maneuver. (**Evaluated**) Platoon will lead the company and be the main effort in the attack. They seize OBJ SADDLE vicinity (**grid**) to allow passage of friendly elements through CACTF. (___) Platoon moves on the company's left and seize OBJ SPUR vicinity (**grid**) to protect the left flank of the main effort. (___) Platoon moves on the company's right and establish a support-by-fire position located vicinity (**grid**).

(2) Fire support. fires are used to obscure movement to OBJ HORSE and to suppress the enemy during the assault. Priority of indirect fires is to (*evaluated*) Platoon, which is allocated one priority target (**120-mm mortar**).

Figure 5-5. Example FRAGO for STX Attack.

```
b. Tasks to Maneuver Unit.
   (1)  (Evaluated) Platoon.
     (a)  Use Passage Point 1.
     (b)  Orient defense from TRP 2 to TRP 3.
     (c)  Be prepared to employ a wire-mined obstacle vicinity TRP 5
during consolidation.
   (2)  (___) Platoon.
     (a)  Use Passage Point 2.
     (b)  Orient defense from TRP 1 to TRP 2.
   (3)  (___) Platoon
     (a)  Use Passage Point 2.
     (b)  Orient defense from TRP 3 to TRP 4.
c. Coordinating Instructions.
   (1)  Company RP is located (grid).
   (2)  Company linkup is (grid).
   (3)  MOPP1 is in effect (date and time).
   (4)  Selective rubbling is authorized. Destruction of an area
larger than one city block requires brigade commander approval.
```

Figure 5-5. Example FRAGO for STX Attack (continued).

4. **Support Requirements**.

a. *Minimum Trainers/Evaluators*: This exercise can be conducted with the company commander or platoon leader acting as the trainer or primary evaluator. At least one more controller or evaluator is required with the OPFOR. Other platoons being trained or evaluated will increase support personnel requirements.

b. *Vehicles/Communications*: All vehicles organic to the platoon should be included. The OPFOR should also have some combat vehicles.

c. *Opposing Force*:. The OPFOR should consist of at least a reinforced squad.

d. *Maneuver Area*: A training area at least 4 by 15 km is needed for cross-country movement and infiltration, and the CACTF is the ideal objective area. The terrain should offer multiple covered and concealed approaches to the objective.

5. **Evaluation**.

Table 5-3 lists the T&EOs from DTMS used to evaluate this STX.

Table 5-3. T&EOs used to evaluate STX Attack.

Task/Function	*Task Number*
Perform Passage of Lines	7-3-1125
Conduct Tactical Movement	7-2-1342
Occupy Assembly Area	7-3-1136
Develop and Communicate a Plan	7-3-1605
Conduct Consolidation and Reorganization (Company/Platoon)	7-2-5027

MECHANIZED INFANTRY PLATOON STX DEFEND

1. **Objective**. This example STX trains collective, leader, and individual tasks for the platoon operation, *Defend Urban Area/Building*.

2. **Interface.** This STX is supported by two warrior drills: *React to Contact* and *Enter and Clear a Building*.

3. **Training.**

a. *Guidance*. FM 7-0 and FM 7-1 provide training guidance, tips, and enhancers for planning and executing STXs. The trainer should review the individual, leader, and collective tasks to be performed during the STX to determine which require initial or refresher training.

b. *General Situation*.

(1) The exercise scenario is as follows: Latest INTSUMs indicate that enemy forces are expected to use a high-speed avenue of approach to counterattack. The approach is dominated by a small urban area containing strongly constructed buildings. The enemy has the capability for indirect fire, CAS, and chemicals. A defense in and around the town is necessary to deny the enemy the use of the avenue of approach. The platoon, as part of a larger force, is ordered to occupy and prepare defensive positions that overlook the avenue of approach. Indirect fire is available.

(2) Conduct of the exercise begins when the platoon receives an OPORD to move to establish a platoon defensive position. An AAR should be conducted after Soldiers enter a building and clear a room, and a final AAR should be conducted once all evaluation notes are compiled. If necessary, portions of the exercise should be repeated until the platoon's performance is satisfactory.

(3) When the platoon receives a warning order to prepare to move to establish a defense, the platoon is part of a company, which is part of a task force, and is located in an assembly area. Figure 5-6 provides a graphic scenario of task performance for the Defend STX. Table 5-4 provides a recommended sequence for performance of T&EOs and the time allotted for each portion of the STX.

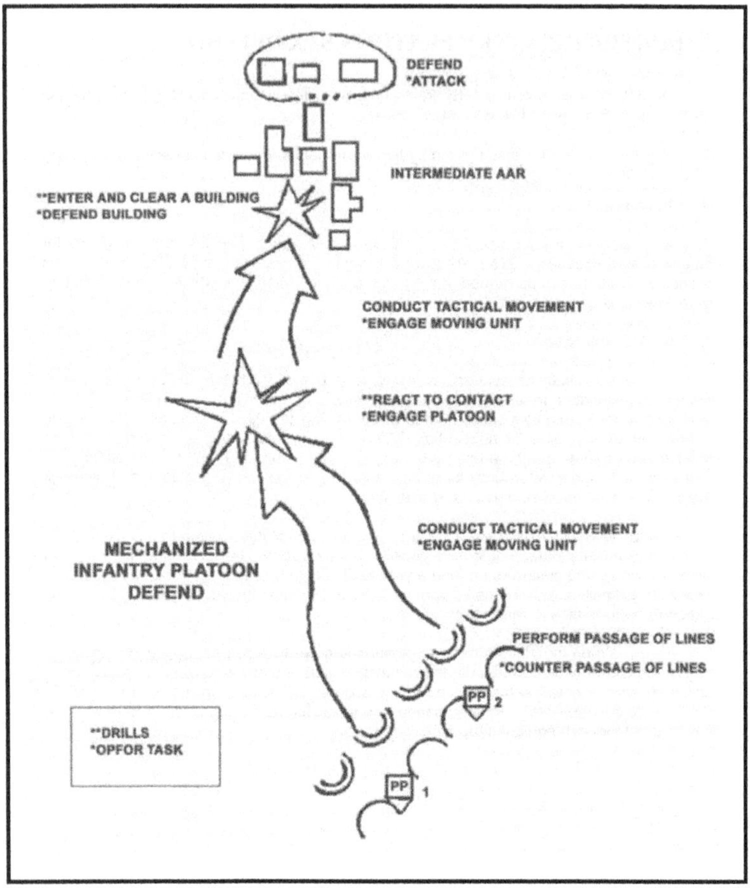

Figure 5-6. Defend STX scenario.

Table 5-4. Sequence of events and times for STX Defend.

Event	*Action*	*Estimated Time*
1	*Perform Passage of Lines*	1 hour
2	*Conduct Tactical Movement*	30 min
3	*React to Contact*	1 hours
4	*Conduct Tactical Movement*	30 min
5	*Enter and Clear a Building (Drill)*	1 hour
6	*Intermediate AAR*	30 min
7	*Defend MOUT/Building*	6 hours
8	*Conduct Final AAR*	1 hour
	***Total Time**	**11 hr 30 min**
*Conducting an exercise at night or in increased MOPP requires additional time.		

c. *Special Situation*. Thirty minutes after receiving the warning order, the platoon receives the company OPORD (Figure 5-7).

(Classification)

Copy no ___ of ___ copies
___ Battalion, ___ Infantry
Unit location or place of issue
Date-time group (with time zone)
Message reference number

OPERATION ORDER NO. _____
References: Map, series no. ___, sheet no. ___, edition ___
Time Zone Used Throughout the Order: ____
TASK ORGANIZATION:
___ **Platoon,** ___ **Platoon(-),** ___ **Platoon, Company Control**

Company HQ
1/1/C-52d Engr
1/2 ___,___
(Reserve)
Company Trains

1. **SITUATION.**

a. **Enemy Forces.**

(1) *Composition, Distribution, and Strength.* We are opposed by elements of the Liberation Guards Militia, Red Brigade. Several platoon-sized outposts are located within the battalion sector and consist of about 30 to 35 Soldiers per platoon. They are equipped with an assortment of Soviet small arms, including AK-47s, RPKs, RPGs, and various wheeled and tracked vehicles including some BMPs. Enemy outposts are at 60 to 70 percent strength and have indirect fire assets available.

(2) *Capabilities.* The enemy units are well entrenched in their defensive positions. The strongpoints have dedicated fire support assets and are capable of using nonpersistent nerve agents to prevent coordinated attacks. No enemy air threat exists but some heavy armored vehicles have been sighted in the area.

(3) *Probable Course of Action.* Intelligence indicates the enemy strongpoints will continue to fortify their positions until relief arrives within 24 hours in the form of a main enemy counterattack.

b. **Friendly Forces.** (__) TF defends in sector NLT (______) along PL (____) to deny enemy forces use of avenue of approach (____). (__) TF must be prepared to continue movement forward of PL (____), on order.

(1) Missions of units on left and right are (as required).

(Classification)

Figure 5-7. Example OPORD for STX Defend.

(Classification)

(1) (__) TF defends in sector NLT (______) from (grid) to (grid) to (grid) to contain any penetrations of the brigade sector.

(1) (___) Battery, (___) Battalion provides DS to 1st Brigade.

c. **Attachments and Detachments**. 1/1/C-52d Engineers attached.

2. **MISSION**. (___) Company blocks enemy forces NLT (date-time group) to prevent them from using high-speed avenues of approach through (name of MOUT site), and to prevent the enemy from counterattacking into the battalion sector.

3. **EXECUTION**.

a. **Concept of the Operation**. (See overlay.) Company prevents the enemy from using the high-speed avenues of approach through (MOUT site) by moving to the site and establishing a blocking position, which overwatches the platoon BPs covering the main avenues. We will block in depth by placing two platoons forward and one back, and maintaining a reserve to block any penetration. We must ensure our sectors of fire are placed to mass fire at the decisive point.

(1) *Maneuver*. (Evaluated) Platoon acts as the main effort, blocking from BP (__), vicinity (grid), to prevent enemy penetration of PL (____). (___) Platoon occupies BP (__), vicinity (grid), to block enemy movement into the main effort's left flank. (___) Platoon occupies BP (__) to prevent the enemy from making a coordinated attack on the main effort.

(2) *Fire Support*. The purpose of fires is to disrupt the enemy when his counterattack reaches PL (____). Priority of indirect fires is to (*evaluated*) Platoon, which is allocated two 155-mm priority targets, one of which may be used as an FPF.

(3) *Intelligence*. Priority of collection is to identify and locate enemy combat reconnaissance patrols and the enemy main effort.

(4) *Engineering*. During movement to the battle position, obstacles are used to ensure mobility. During blocking actions, the purpose of obstacles is to prevent the enemy from advancing on larger streets and to canalize him into more restrictive terrain. Priority of effort is to countermobility. Priority of support is to (*evaluated*) Platoon.

b. **Tasks to Maneuver Units**:

(1) *(Evaluated) Platoon*:

(a) Conduct passage of lines using PP 1.

(b) Lead the company movement after crossing the LD. Be prepared to react to enemy contact when (MOUT site) is reached.

(c) Enter and clear building(s) (number determined by commander).

(Classification)

Figure 5-7. Example OPORD for STX Defend (continued).

(Classification)

(d) Occupy BP (__) and orient on TRPs 2 and 3.
(e) Establish an OP vicinity (grid) to observe main avenue of approach.
(2) *(___) Platoon*:
(a) Conduct passage of lines using PP 2.
(b) Move second in order of movement.
(c) Enter and clear building(s) (number determined by commander).
(d) Occupy BP (__) and orient on TRP 1.
(3) *(___) Platoon*:
(a) Conduct passage of lines using PP 3.
(b) Move last in order of movement.
(c) Enter and clear building(s) (number determined by commander).
(d) Occupy BP (__) and orient on TRPs 4 and 5.
(e) Establish an OP vicinity (grid).

c. **Tasks to Subordinate Units**: Engineer squad travels with the company headquarters behind *(evaluated)* Platoon and supports the main effort's emplacement of obstacles.

d. **Coordinating Instructions**.
(1) Timings.
(a) Brief back ________.
(b) Company rehearsal ________.
(c) SP time ________.
(d) In position NLT (time) ________.
(2) Report all enemy contact.
(3) Report crossing all phase lines.
(4) MOPP status ____; enemy chemical attack is expected.
(5) Air defense status is YELLOW, WEAPONS HOLD.
(6) Leaders briefback the movement and defense plan to the commander at the company CP at (date-time group).
(7) Destruction of subterranean approaches requires brigade commander's approval.
(8) Selective rubbling is authorized. Destruction of more than one building requires commander's approval.
(9) Marking of rooms and buildings is IAW CO SOP.

(Classification)

Figure 5-7. Example OPORD for STX Defend (continued).

(Classification)

(10) Rules of Engagement follow:

(a) Minimize collateral damage to structures.

(b) Civilians have been evacuated. Evacuate to company CP any other civilians encountered.

4. SERVICE SUPPORT. IAW the battalion SOP.

a. **Miscellaneous**.

(1) Battalion trains located vicinity (grid).

(2) Company trains located vicinity (grid).

b. **Materiel and Services**.

(1) *Supply*.

(a) Class I. Ration Cycle M-M-M. Platoons must maintain a three-day supply in each BP.

(b) Class II. Submit requirements for special equipment ASAP. Coordinate with 1SG.

(c) Class III. Refuel all vehicles before (date-time group).

(d) Class IV. Coordinate with S-4 for delivery.

(e) Class V.

- Platoons cache one additional basic load in BP.
- Platoons submit request for additional demolitions NLT (date-time group).

(2) *Services*. Battalion decontamination site is located vicinity (grid).

(3) *Maintenance*.

(a) Priority to (*evaluated*) Platoon M2 BFVs.

(b) UMCP located vicinity (grid).

c. **Medical Evacuation and Hospitalization**.

(1) Battalion aid station located vicinity (grid).

(2) Company CCP located with company CP.

d. **Personnel**.

(1) No replacements for next 72 to 96 hours.

(2) Brigade mortuary affairs point located in BSA.

(3) EPW collection sites located at (grid) for battalion and (grid) for company.

(Classification)

Figure 5-7. Example OPORD for STX Defend (continued).

(Classification)

5. **COMMAND AND SIGNAL**.

a. **Command**. Initially, company CP will be traveling center of company wedge. Location in BP to be determined.

b. **Signal**.
(1) Current SOI in effect.
(2) Primary means of communication in BP is wire, FM, messenger, in that order.
(3) Radio listening silence is in effect until after crossing PL (____) or until contact is made.

(Classification)

Figure 5-7. Example OPORD for STX Defend (continued).

4. **Support Requirements**.

a. *Minimum Trainers/Evaluators*: This exercise is conducted by one company grade officer who is the trainer and primary evaluator.

b. *Vehicles/Communications*: All vehicles organic to the platoon should be included.

c. *Opposing Force*:. The OPFOR should be platoon strength.

d. *Maneuver Area*: A training area at least 2 by 10 km, with hasty fighting positions, is desirable for cross-country movement. The CACTF serves as the objective area. The terrain, ideally, should offer multiple covered and concealed routes. Using terrain that limits the leader to a geographic or predictable route prevents evaluation of the unit's ability to conduct a terrain analysis and select covered and concealed routes.

e. *Barrier Materials.* Sufficient Class IV (sandbags, lumber, and so on) must be made available to fortify positions, IAW Chapter 3, FM 3-06.11.

5. **Evaluation**.
Table 5-5 lists the T&EOs from DTMS used to evaluate this STX, and Table 5-6 lists the supporting tasks.

Table 5-5. T&EOs used to evaluate STX Defend.

Task/Function	*Task Number*
Defend MOUT/Building	7-3-1118
Perform Passage of Lines	7-3-1125
Conduct Tactical Movement (Co/Plt)	7-2-1342

INFANTRY RIFLE COMPANY FTX ATTACK

Table 5-6. Supporting tasks for company FTX.

STX C-II-1	*STX C-II-2*	*STX C-II-3*
Conduct an Infiltration (Co/Plt) Perform Linkup Execute an Assault (MOUT) Conduct Consolidation and Reorganization (Co/Plt)	Perform Cordon and Search Take Action on Contact Conduct an Infiltration (Co/Plt) Perform Linkup Perform Passage of Lines Occupy Assembly Area	Disarm Belligerents Handle Captured Insurgents or Belligerents Apprehend/Detain Noncombatants

1. **Objective**. This FTX is designed to train the Infantry rifle company in conducting offensive operations. This exercise provides practice for the company commander, platoon leaders, squad leaders, and company headquarters personnel in planning, coordinating, and controlling combat operations.

2. **Interface**. This FTX is supported by STX C-II-1, C-II-2, and C-II-3 (Table 5-6). It is also supported by two warrior drills, *React to Contact* and *Break Contact*.

3. **Training**.

a. *Guidance*. FM 7-0 and FM 7-1 provide training guidance, tips, and enhancers for planning and executing STXs. The trainer should review the individual, leader, and collective tasks to be performed during the STX to determine which require initial or refresher training.

b. *General Situation*.

(1) The battalion is ordered to conduct offensive operations in support of the brigade mission. The company is ordered to infiltrate enemy lines and to destroy an insurgent operations base located in the CACTF. The company must be prepared to remain in the CACTF to conduct cordon and search missions.

(2) This exercise should be conducted under various conditions.

(3) When the company receives the battalion warning order the planning process for the attack mission begins. The exercise ends when the company completes a passage of lines, occupies an assembly area, and completes preparations to continue its mission. Figure 5-8 shows a graphic scenario of task performance for the FTX Attack. Table 5-7 provides a recommended sequence for performance of T&EOs and the time allotted for each portion of the STX, including AARs.

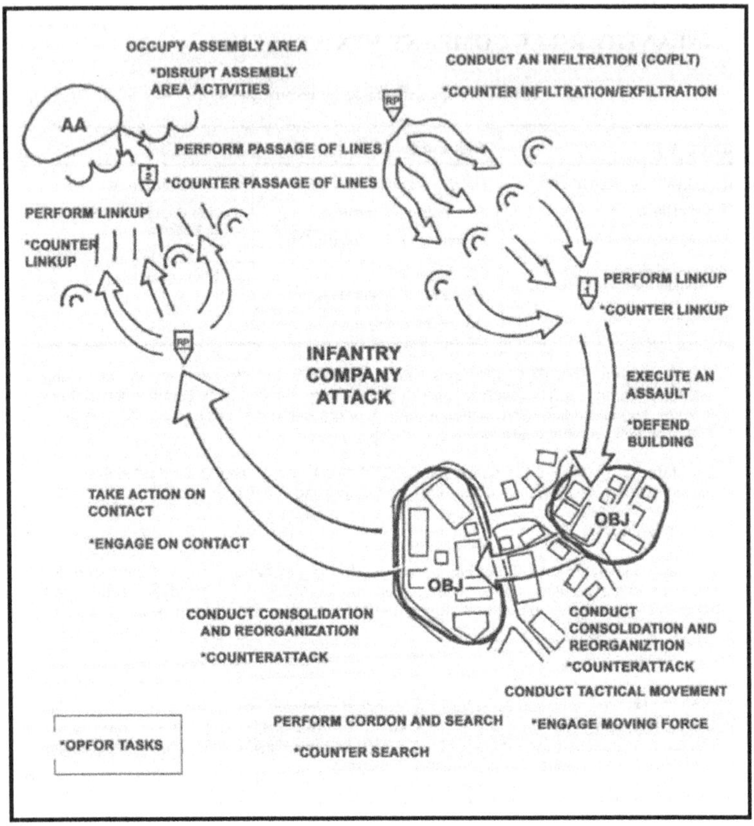

Figure 5-8. Attack FTX scenario.

Table 5-7. Sequence of events and times for FTX Attack.

Event	*Action*	*Estimated Time*
START STX C-II-1		
1	*Issue Company OPORD*	4 hours
2	*Conduct an Infiltration (Co/Plt)*	6 hours
3	*Perform Linkup*	3 hours
4	*Execute an Assault (MOUT) AAR*	3 hours
5	*Conduct Consolidation and Reorganization (Co/Plt)*	3 hours
ISSUE FRAGO, START STX C-II-2		
6	*Perform Cordon and Search*	6 hours
7	*Take Action on Contact / AAR*	3 hours
8	*Conduct an Infiltration (Co/Plt)*	6 hours
9	*Perform Linkup*	3 hours
10	*Perform Passage of Lines*	2 hours
11	*Occupy Assembly Area*	3 hours
START STX C-II-3		
12	*Disarm Belligerents*	3 hours
13	*Handle Captured Insurgents or Belligerents*	4 hours
14	*Apprehend/Detain Noncombatants*	4 hours
15	*Conduct Final AAR*	2 hours
	***Total Time**	**55 hours**

The time required to train an event varies based on the factors of METT-T and on the unit's training proficiency. Regardless, training is conducted to standards, not to time limitations. The following tasks are integrated and evaluated throughout the exercise:

- *Prepare for combat*
- *Maintain operations security*
- *Conduct tactical movement*
- *Perform consolidation*
- *Perform reorganization*
- *Perform stability and support operations*

* Time for intermediate AARs is excluded from this total; additional time is required if much of the exercise is conducted at night or during other limited visibility.

(4) The battalion commander issues an OPORD to the company commanders (Figure 5-9).

(Classification)

Copy no ___ of ___ copies
___ Battalion, ___ Infantry
Unit location or place of issue
Date-time group (with time zone)
Message reference number

OPERATION ORDER NO. ______
References: Map, series no. ___, sheet no. ___, edition ___
Time Zone Used Throughout the Order: ____

TASK ORGANIZATION:
___Company ___Company ___Company Battalion Control
Scout Platoon
81-mm Mortar Platoon
Antitank Platoon
1/C/58th ENGR (DS)
Battalion Trains

1. SITUATION.

a. **Enemy Forces**. Annex A (See intelligence overlay [prepared by brigade S-2].)

(1) *Disposition, Composition, and Strength*. The battalion is opposed by one company from the Dona Anna Militia Guards Battalion. This enemy company consists of 3 platoons with 35 to 40 Soldiers, equipped with M16s, AK-47s, and LMGs. The company headquarters is currently located in the town hall of CACTF. The company is at 90 percent strength and has indirect fire support assets available.

(2) *Capabilities*. The enemy possesses nonpersistent nerve agents and indirect fire assets capable of delivery. He also has some light armored vehicles in the area, but their location is undetermined. No enemy air threat exists. However, the enemy may be able to reinforce CACTF with an additional company, within two hours.

(3) *Most Probable Course of Action*. The enemy defends CACTF and continues using it as a base of operations and to consolidate weapons and material. His intent is to prevent the capture of CACTF and its support base, provided by its facilities and population.

(Classification)

Figure 5-9. Example battalion OPORD for FTX Attack.

(Classification)

b. **Friendly Forces**.

(1) (___) Brigade, (___) Division attacks at (date-time group) to seize the town of CACTF at (grid), to keep the enemy from using it as a base of operations. The intent is to secure the town and the road network in the area, to prevent the enemy from using it as a base of operations.

(2) (___) Battalion, (___) Infantry attacks to seize the road network at OBJ (____), vicinity (grid), at (date-time group) to prevent enemy forces from counterattacking along Route (____).

(3) (___) Battalion, (___) Infantry attacks at (date-time group), to seize the high ground at OBJ (____), vicinity (grid), and to establish defensive positions to block enemy routes in and out of CACTF.

2. **MISSION**. (___) Battalion, (___) Infantry attacks at (date-time group) to seize CACTF, vicinity (grid), NLT (date-time group), to prevent the enemy from using the town as an operations and logistics base.

3. **EXECUTION**.

a. **Concept of the Operation**. The battalion will accomplish its mission, which is to deny the enemy use of CACTF and its surrounding terrain and population, by seizing the town and controlling the key areas in the vicinity. We will conduct a feint in the west and then attack the decisive point within CACTF (OBJ [____]).

(1) *Maneuver*. (___) Company conducts a feint to deceive the enemy, then the main effort (*evaluated*) company seizes OBJ (____), to prevent the enemy from using CACTF as an operational and logistical base. (___) Company supports the main effort by seizing OBJ (____), in order to prevent the enemy from counterattacking the main effort. (___) Company seizes OBJ (____) to support the main effort's attack.

(2) *Fire Support*. Annex C. (See fire support.) Purpose of fires is to suppress the enemy and to obscure movement to CACTF. Priority of artillery fires is to (feinting) Company; on order, priority will shift to (supporting effort) Company. Priority of 81-mm mortar fires is to (*evaluated*) Company.

(3) *Engineering*. The purpose of the engineers is to breach obstacles during the attack, to help (*evaluated*) Company seize OBJ (____). Priority of support is to (*evaluated*) Company, (___) Company, and (___) Company in order. The battalion commander must authorize any booby traps, and point and hasty minefields.

(Classification)

Figure 5-9. Example battalion OPORD for FTX Attack (continued).

(Classification)

b. **Tasks to Maneuver Units**.

(1) *(Evaluated) Company*.

(a) Infiltrate using infiltration lane TANGO.

(b) Seize OBJ (____), vicinity (grid), NLT (date-time group).

(c) Be prepared to perform cordon and search of CACTF, after consolidation and reorganization.

(2) *(Supporting) Company*.

(a) Follow (main attack) Company and infiltrate on infiltration lane TANGO.

(b) Seize OBJ (____), vicinity (grid), NLT (date-time group).

(c) Coordinate the attack with *(evaluated)* Company, so that both units attack at the same time.

(d) Be prepared to perform cordon and search in the objective area after consolidation and reorganization.

(e) Be prepared to assume the mission of *(evaluated)* Company.

(3) *(Supporting) Company*.

(a) Infiltrate using infiltration lane ZEBRA.

(b) Conduct feint to OBJ (____), vicinity (grid), NLT (date-time group).

(c) Be prepared to establish defensive positions vicinity (grid) to prevent enemy movement to and from CACTF.

c. **Tasks to Subordinate Units**.

(1) *(Scout) Platoon*.

(a) During movement, determine location, strength, and disposition of enemy forces in CACTF.

(b) On order, screen the battalion flank from (grid) to (grid) to warn of the approach of enemy units from the southwest

(2) *Mortar Platoon*. Priority of mortar fire is to *(evaluated)* Company. Mortar platoon moves with *(evaluated)* Company, establishes a firing position vicinity (grid), and displaces on order.

(3) *Antitank Platoon*.

(a) Initially, over watch from (grid) to (grid) to secure the unit infiltration.

(b) On order, move vicinity (grid) to support *(evaluated)* Company's consolidation and reorganization.

(Classification)

Figure 5-9. Example battalion OPORD for FTX Attack (continued).

(Classification)

(4) *1/C/58 Engineers.*

(a) Move with *(evaluated)* Company to provide mobility support.

(b) Be prepared to reduce obstacles in CACTF during cordon and search mission.

d. **Coordinating Instructions.**

(1) Report when ready to execute an assault.

(2) Report any movement of enemy forces from the southwest.

(3) No rubbling is authorized, so minimize collateral damage to property.

(4) MOPP0 is in effect.

(5) Terrorist threat warning is red.

(6) Commanders brief back OPORD to battalion commander at battalion tactical CP at (date-time group).

(7) Rules of Engagement (ROE).

(a) Commanders take all steps necessary and appropriate for their units' protection.

(b) The minimum necessary force is used to control the situation.

(c) Leaders must take measures to minimize risk to civilians, without endangering the unit.

(d) Fire is returned directly to its source, not sprayed into a general area.

(e) Firing ceases when the threat is over.

(f) Anyone trying to surrender is allowed to do so.

(g) Civilians and property are treated with respect.

(h) WP can be used vicinity CACTF to aid in isolating the objectives. The requests for indirect fire, within the town, must be authorized by the battalion commander.

4. **SERVICE SUPPORT.** (See Annex D.)

5. **COMMAND AND SIGNAL.**

a. **Command.**

(1) The tactical CP moves with *(evaluated)* Company.

(2) The main CP is vicinity (grid) and displaces on order.

(3) The second in command remains in the battalion main CP.

(4) The rear CP is the alternate battalion main CP.

(Classification)

Figure 5-9. Example battalion OPORD for FTX Attack (continued).

(Classification)

b. **Signal**.
(1) Current SOI is in effect.
(2) Radio-listening silence is in effect for all elements except scouts, until prepared to attack.

ACKNOWLEDGE:
JONES
LTC
OFFICIAL:
SMITH
S-3
Annexes: A–Intelligence
B–Operations Overlay
C–Fire Support
D–Service Support

(Classification)

Figure 5-9. Example battalion OPORD for FTX Attack (continued).

(5) The battalion has destroyed the enemy operating in CACTF, and has seized the town. Two companies are defending against a limited counterattack. One company is in a blocking position. The (*evaluated*) company receives a FRAGO (Figure 5-10) to conduct a cordon and search operation. The purpose of the operation is to locate enemy personnel and equipment remaining in the CACTF.

Enemy personnel in squad-sized elements are reported by the civilian population to be hiding in the CACTF. Reports also indicate numerous weapons caches are being hidden by civilians collaborating with the enemy. Battalion has completed its mission and is preparing to conduct a cordon and search.

After completing this, the battalion exfiltrates to perform a passage of friendly lines. (Evaluated) Company moves to perform cordon and search vicinity OBJ (____), beginning (date-time group) to (date-time group), to locate and seize enemy personnel or equipment.

The (*evaluated*) company notifies the commander of the results of the search, and then notifies the commander when they (*evaluated* company) are ready to start movement. The company exfiltrates on order along Route Gold to perform linkup at (grid) and passage of lines at (grid). (See overlay for lanes.)

Figure 5-10. Example FRAGO for FTX Attack.

4. **Support Requirements**.

a. *Minimum Trainers/Evaluators*:. The company commander acts as the trainer-evaluator for the platoons. The battalion may direct the FTX internal evaluation along with a company evaluator and an FO evaluator. The platoon leaders and platoon sergeants are the primary trainers during the initial phases of the training. The company commander, the XO, the platoon leaders, and the platoon sergeants all act as trainers during the company phases of training. If the exercise is conducted by the battalion, either the battalion commander, XO, or S-3 acts as the trainer and evaluator, but additional battalion assets may be required to support the evaluation phase. The controllers or evaluators are positioned at critical locations throughout the exercise to observe the action. They must not interfere with the conduct of the operation, except to halt it when a safety violation occurs or to conduct an AAR.

b. *Vehicles/Communications*: Those organic to the company are used in the exercise. This includes one designated medical evacuation vehicle per platoon and at least one aidman on site.

c. *Opposing Force*:. The OPFOR ground force should be at least platoon-size. FTXs conducted by the company require internal resourcing; units should rotate to ensure total training. OPFOR tasks and standards are provided in each T&EO.

d. *Maneuver Area*: A training area 5 by 10 km which includes the CACTF is best for this exercise.

5. Evaluation.

Table 5-8 lists the T&EOs used to evaluate this STX.

Table 5-8. T&EOs used to evaluate STX Attack.

Task/Function	*Task Number*
Perform Cordon and Search	7-3-1105
Take Action on Contact	7-4-1107
Perform Passage of Lines	7-4-1125
Perform Linkup	7-4-1128
Conduct Tactical Movement	7-2-1342
Occupy Assembly Area	7-4-1136
Maintain Operations Security	7-4-1409
Prepare for Combat	7-4-1606
Conduct Consolidation and Reorganization (Co/Plt)	7-2-5027
Perform Cordon and Search	7-2-1105
Execute an Assault (MOUT)	7-2-1109
Perform Exfiltration	7-1-1146
Perform Infiltration	7-1-1147
Perform Reorganization	7-1-1608
Perform Consolidation	7-1-1609

MECHANIZED INFANTRY COMPANY OR TEAM FTX DEFEND

1. **Objective**. This FTX is designed to train the mechanized Infantry company or team in conducting defensive operations in urban terrain. It also provides platoons an opportunity to prepare for full-scale FTXs conducted and evaluated by the parent battalion. This exercise provides practice for the company commander, platoon leaders, squad leaders, and company headquarters personnel in planning, coordinating, and controlling combat operations.

2. **Interface**. This FTX is supported by the STX C-II-4, C-II-5, and C-II-6 tasks shown in Table 5-9. It is also supported by two warrior drills, *React to Contact* and *Break Contact*.

Table 5-9. Supporting tasks for company FTX.

STX C-II-4	*STX C-II-5*	*STX C-II-6*
Occupy Assembly Area	Perform Ambush	Operate a Checkpoint
Perform Tactical	Establish a Roadblock/Checkpoint	Interdict Smuggling Operations
Road March	Perform CBRN Operations	Defend Convoy
Perform Relief in Place	Conduct Tactical Movement	
Defend Built-Up Area/ Building	Perform Passage of Lines	
	Occupy Assembly Area	

3. **Training**.

a. *Guidance*. FM 7-0 and FM 7-1 provide training guidance, tips, and enhancers for planning and executing STXs. The trainer should review the individual, leader, and collective tasks to be performed during the STX to determine which require initial or refresher training.

b. *General Situation*.

(1) The battalion is ordered to conduct defensive operations in support of the brigade mission; the company is ordered to perform a relief operation and to prepare a defense in an urban area. The company remains in position for at least 36 hours before receiving a follow-on mission. Most civilians have been evacuated from the area, but a few remain. A terrorist threat exists as well as the possibility of enemy battalion-level offensive actions.

(2) This exercise should be conducted under various environmental conditions. The company also conducts operations under threat of CBRN attack.

c. *Special Situation.*

(1) The planning process for the defense mission begins when the company receives a warning order. The exercise ends when the company completes the passage of lines and occupies an assembly area. Figure 5-11 shows a graphic scenario of task performance for the FTX Defend.

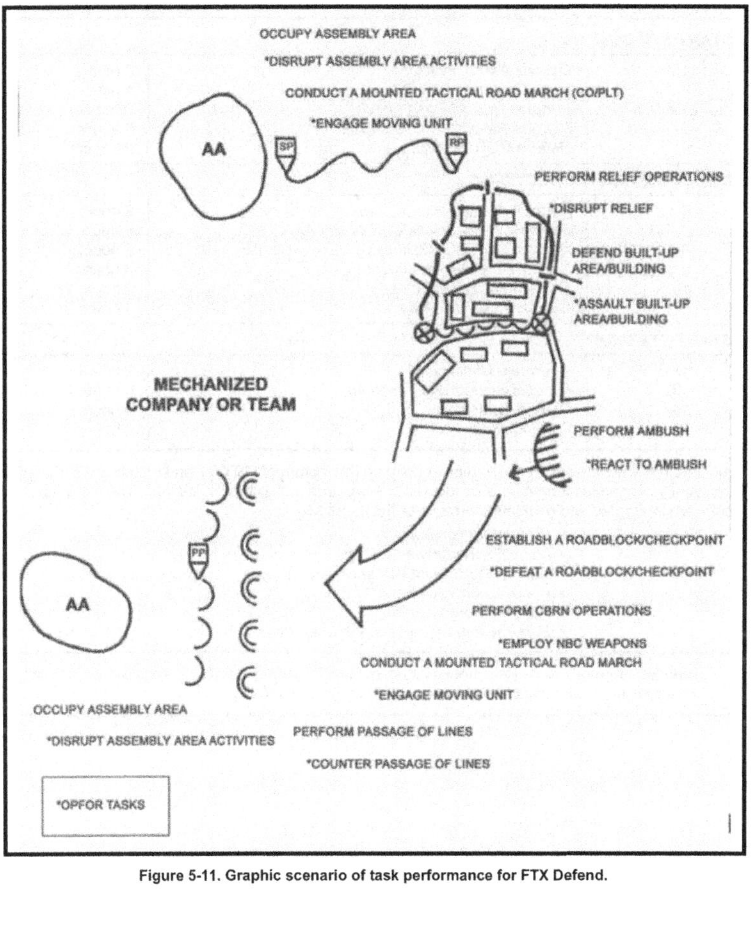

Figure 5-11. Graphic scenario of task performance for FTX Defend.

(2) Table 5-10 provides a recommended sequence for performance of T&EOs from DTMS and the time allotted for each portion of the STX, including AARs.

Table 5-10. Sequence of events and times for FTX Defend.

Event	Action	Estimated Time
START STX C-II-4		
1.	*Occupy Assembly Area*	2 hours
2.	*Issue Company OPORD*	4 hours
3.	*Perform Tactical Road March*	3 hours
4.	*Perform Relief Operations AAR*	6 hours
5.	*Defend MOUT/Building*	24 hours
ISSUE FRAGO, START STX C-II-5		
6.	*Perform Ambush*	6 hours
7.	*Establish a Roadblock/Checkpoint AAR*	6 hours
8.	*Perform CBRN Operations*	6 hours
9.	*Conduct Tactical Movement*	3 hours
10.	*Perform Passage of Lines*	2 hours
11.	*Occupy Assembly Area*	3 hours
START STX C-II-6		
12.	*Operate a Checkpoint*	2 hours
13.	*Interdict Smuggling Operations*	6 hours
14.	*Defend Convoy*	4 hours
15.	*Conduct Final AAR*	2 hours
The time required to train an event varies based on the factors of METT-T and on the unit's training proficiency. Regardless, training is conducted to standards, not to time limitations. The following tasks are integrated and evaluated throughout the exercise: ▪ *Prepare for combat* ▪ *Maintain operations security* ▪ *Conduct tactical movement* ▪ *Perform consolidation* ▪ *Perform reorganization* ▪ *Perform stability and support operations*		
* Time for intermediate AARs is not included in this total; additional time is required if much of the exercise is conducted at night or during other limited visibility.		

(3) The battalion commander issues the defense OPORD to the company (Figure 5-12).

```
                         ____________
                        (Classification)

                                        Copy no ___ of ___ copies
                                      ___ Battalion, ___ Infantry
                                    Unit location or place of issue
                                   Date-time group (with time zone)
                                           Message reference number
OPERATION ORDER NO. _____

Reference: Map, series no. ___, sheet no. ___, edition ___
Time Zone Used Throughout the Order: ____
TASK ORGANIZATION:
___Co ___Tm ___Tm (-) Battalion Control
 1/__/__/AR 2/__/__/AR Scout Platoon
    120-mm Mortar Platoon
Team Tank E Co(-) 1/C/1-222 ADA (Stinger) (DS)
1/__/__/Inf  C/58 ENGR (DS)
2/__/__/Inf  1/1 (GSR) TM A/58 MI Bn
    ATK Force B (1/_/_/Inf)
    Battalion Trains

1. SITUATION.

   a. Enemy Forces. Annex A (see intelligence overlay [prepared by
brigade S-2]).
       (1)  Composition, Disposition, Strength. The company is opposed
by the 1st Dragoons Regiment, 60th Guards Division. This regiment is
at 85 percent strength and has received engineer and artillery
reinforcements. The enemy is equipped with an assortment of former
Soviet and Warsaw pact small arms, machine guns, and armored vehicles
including T-62 tanks.
       (2)  Capabilities. The enemy has established defensive
positions that follow a line from vicinity (grid) to vicinity (grid).
Enemy armored reconnaissance units are operating in the more open
terrain to the southwest of CACTF. The enemy has been observed moving
forces in depth throughout the area in preparation for an offensive
action.
       (3)  Probable Course of Action. The enemy's most probable
course of action will be to begin offensive actions within the next 24
to 36 hours. Enemy encountered includes numerous reconnaissance
elements trying to find the company's vulnerabilities and a main body
trying to exploit any weakness.

                         ____________
                        (Classification)
```

Figure 5-12. Example battalion OPORD for mechanized company or team STX Defend.

(Classification)

b. **Friendly Forces**.

(1) (___) Brigade, (___) Division relieves (___) Brigade in place at (date-time group), to defend in sector from (grid) to (grid) to (grid) to (grid), in order to secure the corps counterattack axis of advance, (grid). The intent is to secure the road network around CACTF to prevent the enemy from impeding the movement of the (___) Brigade, (___) Division.

(2) TF (__) is the main effort in the defense and, NLT (date-time group), holds key terrain vicinity (grid) on our left, to prevent an enemy bypass of the brigade left flank.

(3) TF (__), on our right, defends along Route 360, vicinity (grid) to (grid) to (grid) to (grid), NLT (date-time group), to prevent an envelopment of the right flank. They conduct a supporting attack at (date-time group) to seize the high ground at OBJ OAK, vicinity (grid), to establish defensive positions that prevent the enemy armored reconnaissance units from interfering with the seizure of the pass at (grid) or the attack of the (___) Battalion, (___) Infantry, Mechanized.

2. **MISSION**. TF (____) defends NLT (date-time group), to retain the town of CACTF and its surrounding road network and allow brigade to prevent enemy interference with corps counterattack.

3. **EXECUTION**.

a. **Concept of the Operation**. Annex B (see operations overlay).

(1) Maneuver. Company (main effort) retains the main intersection (traffic loop) in CACTF, to prevent enemy movement through the sector. Team (*evaluated*) blocks movement west of 3d Street, to prevent the enemy from enveloping the main effort. The counterattack force is committed as needed. Team (__) blocks movement east of Oak Street, to canalize the enemy into EA (____). Team (__) defends from BP1, to prevent a coordinated attack against the main effort. E Company occupies BPs, to destroy enemy vehicles and protect the TF flanks.

(2) Fire Support. (See Annex C.) Fires are used to suppress the enemy and obscure our movement during the relief operation. Once positions are occupied, fires are used to suppress enemy overwatch positions, disrupt attempts to breach obstacles, and separate enemy Infantry from their supporting carriers. Priority of CAS and indirect fire is initially to Team (*evaluated*), during the relief operation, and then to Team (main effort). Teams (main effort) and (*evaluated*) each have two priority 155-mm targets. Each team may make one priority target an FPF.

(Classification)

Figure 5-12. Example battalion OPORD for mechanized company or team STX Defend (continued).

(Classification)

(3) Counter-Air Operations. Priority of protection is to the TF trains and main CP. Air defense weapons status is free; air defense warning status is yellow.

(4) Intelligence. Priority of collection is to identify and locate the C2 elements of the MRR and MRB, enemy tank battalions, and the enemy assault detachments' actual axes of advance.

(5) Engineering. The purpose of obstacles is to turn the enemy into restrictive terrain and engagement areas. Priority of support is to Team (main effort), Team *(evaluated)*, Team (__), and Team (__), in that order.

b. **Tasks to Maneuver Units**.

(1) Team (Main Effort).

(a) Sector of defense includes the east side of Front Street, the south side of 1st Street, the west side of 3d Avenue area, and the north side of 3d Street.

(b) Team (main effort) relieves Team (__), TF (__).

(2) Team (Evaluated).

(a) Sector includes the east side of 3d Avenue, the south side of 1st Street, the west side of 6th Avenue, and the north side of 3d Street.

(b) Team *(evaluated)* relieves Team (__), TF (__).

(3) Team (__).

(a) Sector includes the east side of Oak Street, the south side of 1st Street, the east side of Front Street, and the north side of 3d Street.

(b) Team (__) relieves Team (__), TF (__).

(c) Team (__) coordinates directly with right flank team of TF (__), to ensure mutually supporting fires.

(4) Team (__)

(a) Occupy BP 1 vicinity (grid)

(b) Be prepared to occupy BP 2 (grid) to engage enemy in EA (____).

(5) Company E.

(a) Occupy BP 2-1 (grid) and move on order to BP 2-2 (grid).

(b) Provide antiarmor fires to destroy enemy tanks in EA (____).

(6) Counterattack B (Reserve).

(a) Occupy AA Black, vicinity (grid).

(b) Reconnoiter routes into TF (main effort) sector.

(Classification)

Figure 5-12. Example battalion OPORD for mechanized company or team STX Defend (continued).

(Classification)

(3) Counter-Air Operations. Priority of protection is to the TF trains and main CP. Air defense weapons status is free; air defense warning status is yellow.

(4) Intelligence. Priority of collection is to identify and locate the C2 elements of the MRR and MRB, enemy tank battalions, and the enemy assault detachments' actual axes of advance.

(5) Engineering. The purpose of obstacles is to turn the enemy into restrictive terrain and engagement areas. Priority of support is to Team (main effort), Team (*evaluated*), Team (__), and Team (__), in that order.

b. **Tasks to Maneuver Units.**

(1) Team (Main Effort).

(a) Sector of defense includes the east side of Front Street, the south side of 1st Street, the west side of 3d Avenue area, and the north side of 3d Street.

(b) Team (main effort) relieves Team (__), TF (__).

(2) Team (Evaluated).

(a) Sector includes the east side of 3d Avenue, the south side of 1st Street, the west side of 6th Avenue, and the north side of 3d Street.

(b) Team (*evaluated*) relieves Team (__), TF (__).

(3) Team (__).

(a) Sector includes the east side of Oak Street, the south side of 1st Street, the east side of Front Street, and the north side of 3d Street.

(b) Team (__) relieves Team (__), TF (__).

(c) Team (__) coordinates directly with right flank team of TF (__), to ensure mutually supporting fires.

(4) Team (__)

(a) Occupy BP 1 vicinity (grid)

(b) Be prepared to occupy BP 2 (grid) to engage enemy in EA (____).

(5) Company E.

(a) Occupy BP 2-1 (grid) and move on order to BP 2-2 (grid).

(b) Provide antiarmor fires to destroy enemy tanks in EA (____).

(6) Counterattack B (Reserve).

(a) Occupy AA Black, vicinity (grid).

(b) Reconnoiter routes into TF (main effort) sector.

(Classification)

Figure 5-12. Example battalion OPORD for mechanized company or team STX Defend (continued).

(Classification)

(c) Priority for planning:
- Battle position to reinforce main effort.
- Battle position to counterattack into Team (main effort) sector.
- Battle position to counterattack into Team (*evaluated*) sector.

(1) Scout Platoon.

(a) Determine location, strength, and disposition of enemy forces.

(b) Perform a screen of the TF's right flank from (grid) to (grid) beginning (date time group) to prevent enemy infiltration of CACTF.

c. **Tasks to Subordinate Units**.

(1) Mortar Platoon.

(a) Priority of fire to TF (main effort). Move with TF (main effort) and establish firing position vicinity (grid).

(b) Displace on order to (grid).

(2) 1/C/1 222 ADA. Coordinate locations with S-3 NLT (date time group).

(3) C/58 Engineers.

(a) Coordinate the engineer effort with the team commanders NLT (date time group).

(b) Provide an obstacle clearing team to Counterattack Force B, NLT (date time group).

(4) 1/1 (GSR) TM A/58 MI. Coordinate location of assets with the task force S-3 NLT (date time group).

d. **Coordinating Instructions**.

(1) Selective rubbling is authorized, but destruction of more than one building requires TF commander approval.

(2) Destruction of subterranean approaches requires TF commander approval.

(3) Commanders brief back OPORD to TF commander at the TF tactical CP, vicinity (grid), at (date time group).

(4) MOPP1 is in effect as of (date time group).

(5) Decontamination site established vicinity Miller Pond (grid).

(Classification)

Figure 5-12. Example battalion OPORD for mechanized company or team STX Defend (continued).

(Classification)

4. SERVICE SUPPORT. (See Annex D.)

a. **General**.
(1) Combat trains located vicinity (grid).
(2) Field trains located in the BSA, vicinity (grid).

b. **Materiel and Services**.
(1) Supply.
(a) Class I. Ration cycle C-A-C. Team establishes MRE cache sufficient for 48 hours in sector.
(b) Class III. Resupply available on request. Class III located in combat trains.
(c) Class IV. Materials are allocated as shown in Annex D.
(d) Class V. Each team establishes ammunition caches in sector, but should also have two basic loads of ammunition on hand.
(2) Transportation. The main streets are capable of two way heavy vehicle traffic (class 60). Secondary streets may be one way and of lighter construction (class 50).
(3) Maintenance.
(a) Priority by unit: Team (main effort), Team (*evaluated*), Team (__), Company E, Team (__).
(b) Priority by vehicle: M1A1, M2/M3, M113.

c. **Medical Evacuation and Hospitalization**.
(1) TF aid station located vicinity (grid).
(2) Civilian casualties should be transported by nonmilitary means to clinic, located vicinity (grid).

d. Personnel.
(1) No replacements for the next 96 hours.
(2) EPW collection point located vicinity (grid).
(3) Mortuary affairs point located in BSA, vicinity (grid).

e. **Civil-Military Cooperation**.
(1) Curfew in effect 1800 to 0800 daily.
(2) Majority of civilians have been evacuated. Those remaining have been notified of the risk involved.

(Classification)

Figure 5-12. Example battalion OPORD for mechanized company or team STX Defend (continued).

(Classification)

5. COMMAND AND SIGNAL.

a. **Command**.
(1) Brigade main CP located vicinity (grid).
(2) TF tactical CP located vicinity (grid).
(3) TF main CP located vicinity (grid).
(4) TF commander located at tactical CP.
(5) TF XO located at main CP.

b. **Signal**.
(1) SOI index 1 9 in effect.
(2) Primary means of communication is wire, FM, messenger in that order.
(3) Building and obstacle marking IAW TF SOP.

ACKNOWLEDGE:

WILLIAMS
LTC

OFFICIAL:

SMITH
S-3

Annexes: A-Intelligence
B-Operations Overlay
C-Fire Support
D-Service Support

(Classification)

Figure 5-12. Example battalion OPORD for mechanized company or team STX Defend (continued).

(3) The TF has retained the north half of the CACTF after the enemy attack. The enemy was defeated but has established squad-sized to platoon-sized strongpoints in the southern half of the CACTF. The enemy is expected to receive reinforcements within the next 12 to 24 hours. Theprevent enemy reinforcements from moving to the strongpoints to strengthen and resupply them, TF headquarters orders the team to conduct platoon-sized ambushes. They also order the team to prepare to establish a roadblock, to control refugee flow through the area of operations, before the team moves out of the area. Headquarters issues a FRAGO (Figure 5-13).

```
Numerous enemy strongpoints are located to your front, vicinity (grid), (grid), and (grid). We expect these strongpoints to receive reinforcements within the next 12 to 24 hours.

Your mission is to establish platoon ambushes on the likely avenues of approach to the strongpoints, NLT (date-time group), in order to prevent reinforcements or resupply to these strongpoints.

Notify the TF commander when you are prepared to move.

Coordinate with TF scouts for guides. On order, after return from ambush missions, establish roadblocks at (grid) and (grid). Be prepared to move within the next 36 to 48 hours to conduct further follow-on missions.
```

Figure 5-13. Example FRAGO for FTX Defend.

4. **Support Requirements**.

a. *Minimum Trainers/Evaluators*: This exercise should be conducted for practice with the company commander as the trainer-evaluator for his platoons. The battalion may direct the FTX (internal evaluation) using a company evaluator and an FO evaluator. The platoon leaders and platoon sergeants are the primary trainers during the initial phases of the training. The company commander, the XO, the platoon leaders, and the platoon sergeants act as trainers during the company phases of training. If the exercise is conducted by the battalion, either the battalion commander, the XO, or the S-3 acts as the trainer and evaluator, but may require additional assets from the battalion to support the evaluation phase. The controllers or evaluators must be positioned at critical locations throughout the exercise, to observe the actions. They must not interfere with the conduct of the operation, except to halt it when a safety violation occurs or to conduct an AAR.

b. *Vehicles/Communications*: Those organic to the company are used.

c. *Opposing Force*:. The OPFOR ground force should be at least platoon-size. FTXs conducted by the company require internal resourcing, and units should be rotated to ensure total training. T&EOs provide training tasks and standards for the OPFOR.

d. *Maneuver Area*: A training area at least 5 by 15 km is best for this exercise.

e. *Consolidated Support Requirements*: This exercise requires the items shown in Table 5-11.

f. *Barrier materials.* Sufficient Class IV must be made available to fortify positions, IAW FM 3-06.11.

Table 5-11. Ammunition and pyrotechnics support requirements for a six-day FTX.

Ammunition	*DODAC*	*Per Plt/Co/Bn FTX*
5.56-mm blank	1305A080	120 rounds per rifle
5.56-mm SESAMS		120 rounds per rifle
5.56-mm blank	1305A075	200 rounds per M249 LMG
5.56-mm SESAMS		200 rounds per M249 LMG
7.62mm	1305A111	200 rounds per BFV/M1
		400 rounds per M60/240B
ATWESS cartridge	1370L367	6 (Viper) (for LAW) 18 per Dragon 14 per TOW system 9 per OH-58D/AH-64
Pyrotechnic device	1370L602	6 per M1 2 per target
Hand grenade fuze (M228)	1330G878	150 per battalion
Stun grenade M84	1330GG09	150 per battalion
Smoke grenade, HC		150 per battalion
Smoke pot		6 per battalion

5. **Evaluation**.

Table 5-12 lists the DTMS T&EOs used to evaluate this STX.

Table 5-12. T&EOs used to evaluate STX Attack.

Task/Function	*Task Number*
Defend MOUT/Building	7-3/4-1118
Conduct a Dismounted Tactical Road March (Co/Plt)	7-2-1189
Conduct a Mounted Tactical Road March (Co/Plt)	7-2-1198
Perform Relief Operations	7-3/4-1124
Perform Passage of Lines	7-3/4-1125
Conduct Tactical Movement	7-2-1342
Occupy Assembly Area	7-3/4-1136
Establish a Roadblock/Checkpoint	7-3/4-1401
Perform CBRN Operations	7-3/4-1406
Maintain Operations Security	7-3/4-1409
Prepare for Combat	7-3/4-1606
Conduct Consolidation and Reorganization	7-2-5027
Perform Ambush	7-2-1106

TARGETRY

5-10. Table 5-13 lists the combination of precision and nonprecision targets to support the diversity of scenarios required to support combined arms and branch specific lane training. Precision targets, less than 50 meters, are used in and around buildings within the CACTF and reinforce precision marksmanship techniques described in FM 3-22.9 (Table 5-13).

Table 5-13. CACTF target requirements.

Event	*Number Of Lanes*	*Human Targets*		*Vehicle Targets*	*Totals*	
					HT	*VT*
Company/Platoon STX Offense	3/9	Precision	Nonprecision	9	45	9
		15	30			

CONSIDERATIONS FOR TRAINING WITH ARMOR

5-11. Built-up areas consist mainly of man-made features such as buildings, streets, and subterranean systems. These features of urban terrain create a variety of tactical problems and possibilities. In order to ensure that the tank platoon can operate effectively in the urban environment, the platoon observation and direct-fire plans must address the ground-level or surface fight (in streets and on the ground floor of buildings), the aboveground fight (intrasurface and supersurface), and the subterranean fight (Figure 5-14). Infantry can assist the tank platoon by—

- Locating targets for tanks to engage.
- Destroying antitank weapons.
- Assaulting enemy positions and clearing buildings with tank support.
- Protecting tanks from antitank fires.

Figure 5-14. Urban battlespace.

TRANSPORTING INFANTRY

5-12. At times, the tank platoon must transport Infantrymen (Figure 5-15). They only do so when not expecting contact. If the platoon is moving as part of a larger force, and if it is tasked to provide security for the move, then the lead section or element does not carry Infantry.

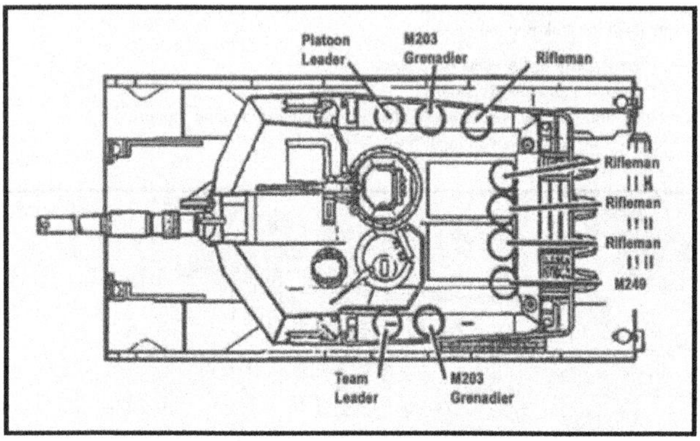

Figure 5-15. Example positions for Infantry riding on a tank.

5-13. Infantry and Armor leaders must observe the procedures, precautions, and considerations in Figure 5-16 when Infantrymen ride on tanks:

- Infantry should thoroughly practice mounting and dismounting procedures and actions on contact.
- Infantry must always alert the TC before mounting or dismounting. They must follow the commands of the TC.
- Infantry platoons should be broken down into squad-sized groups, similar to air assault chalks, with the Infantry platoon leader on the Armor platoon leader's vehicle and the Infantry PSG on the Armor PSG's vehicle.
- Platoon leaders, PSGs, and team leaders should position themselves near the TC's hatch, using the external phone (if available) to talk to the TC and relay signals to the unit.
- If possible, the lead vehicle should not carry Infantrymen. Riders restrict turret movement and are more likely to be injured or killed on initial contact.
- Whenever possible, Infantrymen should mount and dismount over the left front slope of the vehicle. This procedure ensures that the driver can see the Infantrymen and that the Infantrymen DO NOT pass in front of the coax machine gun. Infantrymen must ensure that they remain behind the vehicle's smoke grenade launchers. This will automatically keep them clear of all weapon systems.
- Infantrymen must always have three points of contact with the vehicle; they must watch for low-hanging objects such as tree branches.
- All Infantrymen should wear hearing protection.
- Infantrymen should not ride with anything more than their battle gear. Rucksacks and B-bags should be transported by other means.
- Infantrymen should scan in all directions. They may be able to spot a target the vehicle crew does not see.

Figure 5-16. Precautions for Infantrymen riding on a tank.

5-14. Passengers should be prepared to take the actions on contact shown in Figures 5-17 and Figure 5-18.

- Wait for the vehicle to stop.
- At the TC's command, dismount immediately (one fire team on each side). ***Do not*** move forward of the turret.
- Move at least 5 meters to the sides of the vehicle.
  Do not move behind or forward of the vehicle.
- ***Do not*** move in front of vehicles unless ordered to do so. Main gun discharge overpressure can inflict severe injury or death to forward dismounts.
- ***Do not*** dismount a vehicle unless ordered or given permission to do so.
- ***Do not*** dangle arms or legs, equipment, or anything else off the side of a vehicle; they could get caught in the tracks, causing death, injury, or damage to the equipment or vehicle.
- ***Do not*** place too many riders on the vehicle.
- ***Do not*** fall asleep when riding. The warm engine may induce drowsiness; a fall could be fatal.
- ***Do not*** smoke when mounted on a vehicle.
- ***Do not*** stand near a moving or turning vehicle at any time. Tanks have a deceptively short turning radius.

Figure 5-17. Actions on contact for Infantrymen riding on a tank.

DANGER

1. The overpressure from the tank 120-mm cannon can kill a dismounted Infantryman within a 90-degree arc extending from the muzzle of the gun tube out to 200 meters.

2. From 200 to 1,000 meters along the line of fire, on a frontage of about 400 meters dismounted Infantry must be aware of the danger from discarding sabot petals, which can kill or seriously injure Soldiers.

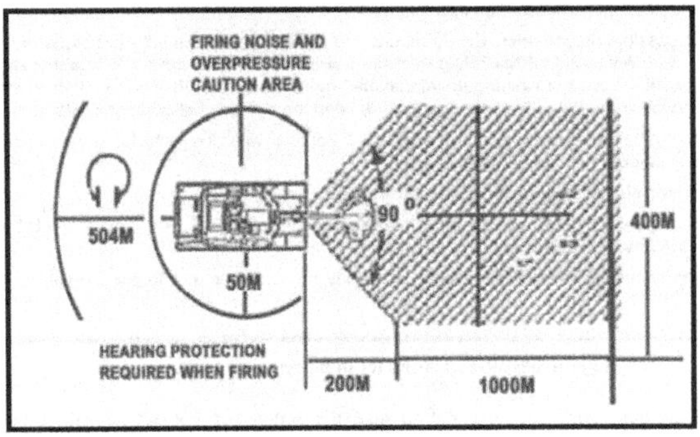

Figure 5-18. Danger areas around a tank when firing the 120-mm main gun.

ADDITIONAL CONSIDERATIONS

5-15. Additional considerations and preparations for transporting Infantrymen include the following:

- Armor--
  - Uses main gun fire to reduce obstacles or entrenched positions for the Infantry.
  - Takes directions from the Infantry ground commander (platoon leader/PSG/squad leader) to support their fire and maneuver.
  - Provides reconnaissance by fire for the Infantry.
  - -- Should know and understand how the Infantry clears buildings, how they mark cleared buildings, casualty evacuation plan, signal methods, engagement criteria for tank main gun, front line trace reporting, ground communication from the tank with the dismounted Infantry.
  - -- Uses its night vision capability to augment and supplement the Infantry's night vision capability.
- Infantry--.
  - -- Provides real time information for the tank crewmen to help them overcome tank noise and the lack of ground situational understanding.
  - -- Provides reconnaissance and fire direction of enemy positions for main gun attack.

- Considerations for dismounted tank security include--
  - -- Tank crewman should rehearse the mounting and dismounting of Infantry from their vehicle, briefing the Infantrymen on safety procedures for the vehicle and weapon systems.
  - -- Tank commanders need to rehearse communicating with dismounted Soldiers by way of TA-1 and DR-8 in the bustle rack.
- Vehicle preparation for combat in urban terrain should cover--
  - -- Keep at least one ballistic shield to the "Dog House" closed. Most engagements will be under boresight range, and the battlesight technique will suffice.
  - -- Place sandbags around antenna connections and electrical wiring on the turret top.
  - -- Place extra coax ammunition inside the turret.
  - -- Remove any highly flammable products from the outside of the vehicle and from the sponson boxes.

VEHICLES, WEAPONS, AND MUNITIONS

5-16. Numerous factors related to vehicles and their organic weapons and munitions affect the tank platoon's urban operation planning and execution, including the following:

- The preferred main gun rounds in the urban environment are HEAT, MPAT (ground mode), and MPAT-OR. These all perform much better than sabot rounds against bunkers and buildings.
- HEAT ammunition will open a larger hole in reinforced concrete or masonry structures than MPAT or MPAT-OR. Both MPAT and MPAT-OR, however, offer greater incapacitation capability inside the structure.
- HEAT ammunition arms approximately 60 feet from the gun muzzle. It loses most of its effectiveness against urban targets at ranges of less than 60 feet.
- MPAT and MPAT-OR rounds arm approximately 100 feet from the muzzle of the gun. Because of the shape and metal components of the projectiles, however, this ammunition remains effective at ranges of less than 100 feet.
- Sabot petals, including those on MPAT and MPAT-OR, endanger accompanying Infantry elements. They create a hazard area extending 70 meters on either side of the gun-target line out to a range of 1 kilometer.
- The tank's main gun can depress only to -10 degrees and can elevate only to +20 degrees. This creates considerable dead space for the crew at the close ranges that are typical in the urban environment.
- The external M2 HB machine gun can deliver a heavy volume of suppressive fire and penetrate light construction, buildings and most barricades. The M2 HB MG can elevate to +36 degrees; however, the TC must be unbuttoned to fire the M2 on the M1A2 or M1A2 SEP.
- The M240 coax machine gun can effectively deliver suppressive fires against enemy personnel and against enemy positions that are behind light cover.
- The loader's M240 machine gun can effectively deliver suppressive fire against enemy personnel and against enemy positions that are behind light cover; however, the loader must be unbuttoned to operate it. This weapon may be dismounted and used in a ground role if units are equipped with the M240 dismount kit.

- When buttoned up, the tank crew has limited visibility to the sides and rear and no visibility to the top. Figures 5-19 and 5-20 show the dead spaces associated with tank operations in an urban environment.
- FM 3-20.12 explains special uses for tank-mounted machine guns in the urban environment.

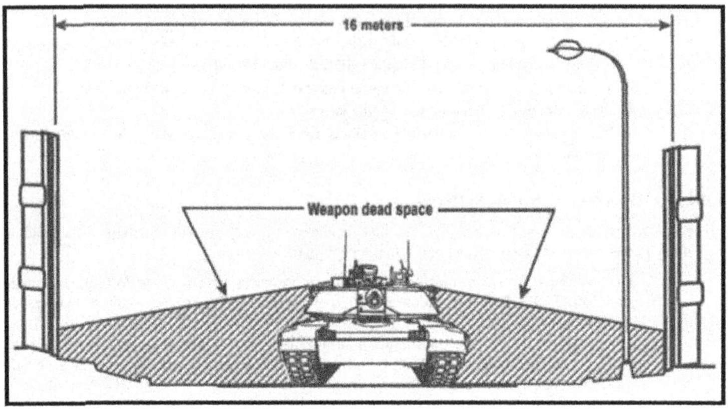

Figure 5-19. Tank weapon dead space at street level.

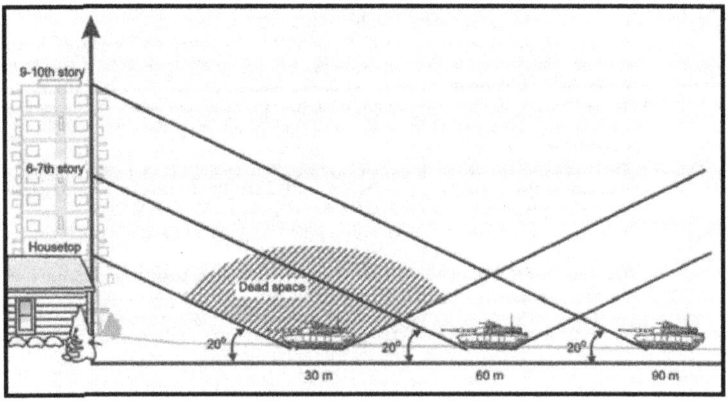

Figure 5-20. Tank main gun and coax dead space above street level.

Tank Platoon Command and Control in Urban Operations

5-17. Communications problems, fire control, proximity and visibility, personnel factors, ROE, ROI, civilians, and slow-pace urban operations all affect the tank platoon's urban operations planning and execution:

Communications Problems

5-18. The low-level task organization that may take place during urban operations requires elements to establish additional communications links, which can be disrupted by buildings and other urban terrain features.

Fire Control

5-19. Extensive direct fire planning and restrictive fire control measures are an absolute requirement in urban operations.

Proximity and Visibility

5-20. Friendly elements often must operate in confined and restrictive areas during urban operations, and they may not be able to see other nearby friendly forces. These factors significantly increase the danger of fratricide.

Personnel Factors

5-21. Urban operations impose significant, and often extreme, physical and psychological demands on Soldiers and leaders.

ROE, ROI, and Civilian Considerations

5-22. The ROE or ROI may restrict the use of certain weapon systems and TTP. As an integral part of urban operations, noncombatants create special operational problems. Thedeal with these concerns, units operating in urban terrain must know how to effectively employ linguists and counterintelligence and civil affairs teams.

Slow Pace Urban Operations

5-23. This operation usually prevents the tank platoon from taking full advantage of the speed and mobility of its tanks.

Appendix A

Training Tips

The effectiveness of urban operations training depends on the imagination and thoroughness applied to a training plan. All trainers, from training managers to junior leaders, should prepare their units to fulfill their urban operations mission. They can employ procedures that are based on training doctrine and proven techniques. This appendix discusses several techniques for conducting urban operations training. Trainers make the choice based on the training needs and available resources.

CONCURRENT TRAINING

A-1. Concurrent training allows groups of Soldiers to train on different tasks at the same time. This training requires a plan for rotating Soldiers between the stations that teach different tasks. This procedure is referred to as round-robin training. Concurrent training is an efficient training technique since it makes the best use of training time and facilities. Concurrent training in the urban operations training facilities works best with performance-oriented training. It also requires detailed planning, control, coordination, and assistant trainers. Each station should conduct training that is relevant to, and supports the training the unit is conducting.

URBAN ASSAULT COURSE AND BREACH FACILITY

A-2. These facilities are well suited for conducting concurrent training. After mastering the tasks at one station, a group of Soldiers rotates to the next station in the sequence. Theavoid bottlenecks at any of the stations, trainers plan instructional blocks that last about the same length of time. Concurrent training is often conducted with dedicated instructors at each station. The *advantage* of this technique is that it ensures uniformity of performance standards and saves on trainer preparation time. The *disadvantage* is that the Soldier is not always trained by his leader. Therefore, when feasible, small-unit leaders should rotate with their Soldiers from station-to-station so they can take part in the training. This technique demands prior planning, preparation, and instructors that can teach.

URBAN ASSAULT COURSE AND COMBINED ARMS COLLECTIVE TRAINING FACILITY

A-3. These facilities are well suited for conducting concurrent exercises. Concurrent exercises are variants of concurrent training. In a concurrent exercise, elements of the same unit receive different missions. These missions correspond to each element's training objective for that exercise. This technique can be used during FTXs, STXs, or CFXs. For example, one unit attacks; a second unit supports by fire; a third unit acts as a reserve. Also, elements of the same unit can challenge each other. For example, if a trainer has been training his unit in urban offense and defense operations, he can create a scenario in which one force is required to attack the other. This technique allows the unit to train with an OPFOR without an extra investment of resources. Leaders must remember to carefully plan the force ratios between the OPFOR and counterattacking forces.

DEMONSTRATIONS

A-4. Trainers must demonstrate the correct way to perform the tasks that the Soldiers will be required to perform. A demonstration is more effective when the trainer uses hand-on techniques. A live demonstration is an effective teaching method since it holds Soldiers' interest and they are able to learn with their eyes. The UAC and the Breach Facility are ideal facilities for teaching individual and collective tasks using demonstrations. The trainer must ensure that each Soldier. Performs the task being taught to guarantee that the soldier has learned. Key points should be explained to the satisfaction of each Soldier before progressing to the next step. The trainers can use video cameras to film the action occurring inside the buildings to show the Soldiers the correct way to perform a task or use the same action to explain a better technique to accomplish the same task. This technique is very effective because the Soldiers can watch the demonstrations on a remote monitor, and trainers can replay the videotape as often as required to reinforce his training objective.

EXERCISES

A-5. Training exercises are used to develop tactical SOPs, teamwork among the participants, and Soldier confidence in his leaders and equipment. Table A-1 matches training exercise types to appropriate training facilities.

Table A-1. Training exercise matrix.

EXERCISE	*CACTF*	*SHOOT HOUSE*	*SIMULATIONS*	*LOCAL TOWN*
MAPEX	X		X	X
TEWT	X			X
CPX	X		X	X
CFX	X			X
BSX			X	
LFX		X		
STX	X	X		X
FTX	X			X

MAP EXERCISE

A-6. A MAPEX allows participants to plan and execute urban operations using maps, overlays, sand tables, photographs, or models. MAPEXs develop teamwork and skills in terrain analysis, planning, coordination, order writing, and employment of urban operation tactics and are more suited for battle staff and leaders.

- Using a model of the CACTF is an excellent training aid for instructing Soldiers and leaders. Gather leaders around the map or model and have them devise a tactical plan. Use figures or markers to represent friendly units and the OPFOR. Create tactical situations that challenge participants to react as they would in real combat.
- Incorporate direct fire, casualties, CBRN, logistics, and civilians into the exercise. Ensure the ROE for each exercise is clear to participants.
- Have plenty of detailed maps available for use in the MAPEX and in the planning and execution of subsequent CFX operations.

Tactical Exercise Without Troops

A-7. A TEWT focuses on leader and staff tasks. The senior trainer selects the terrain for the urban TEWT. Either the CACTF or a real village, town, or city is appropriate. The procedure for conducting a TEWT is as follows:

- Create a scenario and enemy situation, and assign missions.
- Allow the leaders to devise their reconnaissance plans and implement them on the terrain.
- Require the leader to brief subordinates while walking the terrain.
- Ensure their plans include the following tasks:
  - -- Analyze the terrain.
  - -- Employ units according to the terrain analysis.
  - -- Emplace weapons.
  - -- Devise a unit plan.
- Conduct an AAR.

Command Post Exercise and Battle Simulation Exercise

A-8. CPXs and BSXs allow leaders and battle staffs to perform their tasks in a unit CP, where they receive information as reports requiring certain reactions. This type of training can be achieved through the constructive simulations training portion of the urban operations training strategy using JANUS or JCATS. A company CP is usually associated with a battalion level or higher CPX, but a company can conduct its own CPX. The headquarters section can set up a CP in a CACTF building, displace to another building, and repeat the exercise until they meet the required standards. The section can practice hiding wire on pavement, stringing it between buildings, and running it through sewers. The CP can practice receiving reports, posting situation maps, and sending reports forward to higher headquarters. The CPX/BSXs are most effective in achieving the following general training objectives:

-- Preparing plans.
-- Issuing orders.
-- Building teams.
-- Reconnoitering, selecting, and tactically occupying CP locations.
-- Establishing and using communications.
-- Preparing and sending reports.
-- Coordinating between subordinate elements.

Command Field Exercise

A-9. The CFX is used to train commanders, staffs, and junior leaders. The CACTF or a local town can accommodate this type of exercise. However, the unit commander must go through the correct procedures when conducting training off federal property. The CFX employs all of a unit's command and control assets, with a number of subordinate elements reduced according to a set scale. For example, one Soldier might represent an entire squad, one vehicle a platoon, or one gun an entire battery.

Preliminary Training

A-10. Preliminary training, such as CPXs, TEWTs, MAPEXs, and BSXs, should precede the CFX. Units conduct operations, and controllers resolve battle outcomes. The exercise ends with an AAR. Fewer controllers are needed in a CFX than in an FTX. However, if the number of unit functions increases, the number of controllers must increase also.

Planning Procedures

A-11. Combat in urban areas entails centralized planning with decentralized execution.

Centralized

A-12. Centralized planning can be practiced in a CFX; however, it can be practiced more efficiently in a CPX, TEWT, MAPEX, or BSX.

Decentralized

A-13. Decentralized execution means that platoon, squad, and team leaders lead the battle in an urban area. However, because element participation is scaled down in a CFX, trainers usually find STXs, FTXs, and TEWTs more useful.

LIVE-FIRE EXERCISE

A-14. Chapter 3 and local range regulations provide specific guidance for conducting LFXs at the shoot house. Live-fire training allows Soldiers and leaders to maneuver while firing service ammunition or SRTA. An LFX is most productive when preceded by non-live-fire practice (dry fire or dry run). However, dry fire should *always* precede live fire, with either TES or SESAM. An AAR should follow each repetition, and units should only proceed to the next level of difficulty after mastering the current step. The training in an LFX occurs in the following sequence, which can be shortened if time and ammunition dictate.

1. Dry fire to practice unit drills and individual tasks.
2. Conduct an AAR.
3. Practice drills, tasks a second time with limited use of ammunition to show the complex coordination of fire and maneuver required.
4. Conduct an AAR.
5. Practice exercise a third time with full use of ammunition to enhance realism and build confidence.
6. Conduct an AAR.

SITUATIONAL TRAINING EXERCISE

A-15. STXs are mission-oriented, limited exercises used to train a unit to perform a single collective task or a series of related tasks and drills. With their distinct start and stop points, STXs represent a segment of battle. An STX is more flexible and complex than a single T&EO or drill. In fact, an STX typically includes drills, leader tasks, and separate individual tasks, and involves a unit's "slice" of external combat assets. An STX often precedes an FTX or CFX, and can be conducted in the CACTF by creating a brief situation and mission. For example, the MP company commanders might want to teach his Soldiers about straggler control. They can string the tasks together into small scenarios such as STXs, describe a tactical situation that produces stragglers from nearby units, use other Soldiers to represent the stragglers, and train Soldiers to process stragglers at straggler collection and transport points.

FIELD TRAINING EXERCISE

A-16. FTXs allow units to execute collective tasks under conditions that closely resemble actual urban terrain. FTXs are suitable for the CACTF.

- An FTX requires significant time, resources, and planning. The time is well spent if units master their collective tasks. An FTX is an excellent way to conduct the *run* stage of collective training. Additionally, OPFOR and MILES greatly enhance the training value of the FTX.
- A unit's urban operations FTX training plan addresses ROE, maneuver restrictions, and the controller plan. Soldiers must know the exercise's training objectives before they begin, so they can focus on the correct tasks. Every participant, regardless of position or rank, must fully understand all safety requirements.

- To make the most of their training time in the field, leaders must plan in great detail. The concept of multiechelon training demands that they consider what training can occur at each level in each event.

EVALUATORS

A-17. Soldiers from the training unit or from another unit can evaluate urban operations training, especially during the *run* stage of training in an STX, FTX, or CFX. A dedicated evaluator provides objective and thorough comments on the unit's performance throughout all stages of the exercise. Evaluators monitor the critical point of the action but must avoid interfering. They must not disrupt unit performance or undermine realism. They must be trained and given standardized evaluation guides if their feedback is to be meaningful; MTPs and T&EOs are a good start. Evaluators facilitate AARs alone or along with the unit leader.

TACTICAL ENGAGEMENT SYSTEM

A-18. The TES realistically replicates the effects of actual fire in training. It uses laser adapters on the weapons and laser-sensitive equipment on the Soldiers and vehicles. A TES weapon emits a laser beam when fired. Accurate fire strikes the laser-sensitive equipment and scores a hit. The TES requires Soldiers to aim accurately and to use cover and concealment effectively. When used with an OPFOR, TES offers realistic training. It is an excellent training aid for an FTX or STX. It is also effective at the shoot house as a substitute for live fire. The two major shortcomings of TES are that it allows Soldiers to seek cover behind structures that normally would not protect them, and that it does not replicate suppression.

NAVIGATION

A-19. Navigational techniques for urban areas differ from those needed on other terrain.

- A typical 1:50,000-scale military map provides insufficient detail for accurate navigation. Therefore, Soldiers must learn to navigate using commercial maps, sketches, and aerial photographs. Leaders must learn how to plan operations using these items.
- Underground navigation is a special challenge. Soldiers must know subway plans and be able to identify tunnels, elevated tracks, stairways, and other critical features. They must learn how to measure distance, remain oriented, and mark their routes.
- Destruction often erases urban features that could aid in navigation. Soldiers must learn to orient using distinctive, difficult-to-erase features such as bridges, waterways, towers, and intersections.
- Leaders conduct navigational exercises in an actual urban area. Soldiers must move from point to point while controllers at each point monitor and verify Soldiers' progress.
- The GPS is a navigational aid that triangulates with satellites to calculate position. However, this system may be affected by nearby tall buildings, and its effectiveness decreases inside buildings or underground.

LIMITED VISIBILITY TRAINING

A-20. Urban operations involve day and night operations. FTXs, STXs, CFXs, and CPXs are the most valuable exercises for night training.

- Field-expedient lighting, NVDs, illumination, and electrical lighting must be exploited when the tactical situation allows. Soldiers must learn discipline in their use of light, and also learn the lighting techniques that allow them to carry out their urban training program.

- Situational awareness is the Soldier's knowledge of his location, the locations of friendly forces, and the locations of the enemy on the battlefield. Situational awareness is the key to avoiding fratricide, especially during limited visibility operations. Units must plan to use all the equipment and devices available to mark friendly units and to avoid fratricide, without exposing themselves to enemy observation.
- Reverse-cycle training should be part of an urban training program. Limited visibility fighting skills are perishable, but they can be sustained and improved by effective training. The goal is for every Soldier to operate as effectively at night as during the day.

SAFETY AND AMMUNITION CONSIDERATIONS

A-21. When using ammunition and pyrotechnics, all associated safety requirements must be followed. This paragraph describes some of the safety equipment and training procedures to consider:

SPECIAL EFFECT SMALL-ARMS MARKING SYSTEM

A-22. SESAMS is 5.56-mm and 9-mm color-marking training ammunition. It is designed for close combat training. It is used with the close combat mission capability kit (CCMCK) adapter kit, which is installed on the M4/16 rifle, M9/M11 pistol, and M249 SAW. Different colors can be assigned to OPFORs for engagement identification and fratricide tracking. SESAMS has a 30-meter effective range and activates TES (MILES), so the two are used together. Anytime during a SESAMS exercise that a Soldier loses any of his protective gear, he must immediately cover his face and eyes with his hands. He must stay put until the engagement ends, or until he is told that it is safe to replace his safety equipment. Everyone, to include observer-controllers, must wear the protection shown in Table A-2.

Table A-2. SESAMS protection.

SKIN	Wear gloves, and cover all exposed skin. At close ranges, the SESAMS projectile can break skin.
HEAD	Take every precaution to protect your head.
EYES	Keep your eyes completely enclosed within eye protection. You can use standard Army sun, wind, and dust goggles.
EARS	Wear hearing protection to prevent hearing loss.
MOUTH	Keep your mouth covered.

DANGER

Failure to wear the required protection during SESAMS exercises can result in injury or death.

M84 STUN GRENADE

A-23. The hand-deployed M84 stun grenade serves as a reliable, effective, nonlethal means of neutralizing, diverting, and disorienting enemy personnel. It is used tactically as well as in training.

A-24. The M84 produces an intense (1 to 2.5-million candlepower) flash and, at 1.5 meters, a 170 dBs bang. Even though this is too low to produce permanent hearing damage, one-time use Soldiers are required to wear hearing protection anytime noise levels exceed 85 dBs.

A-25. The M84 stun grenade has an M201A1 fuze and a pyrotechnic output charge in a synthetic casing with a one- to two-second delay fuze, which cannot be cooked off. The grenade body has a nonfragmenting metal hexagonal casing, which, once deployed, resists rolling.

SMOKE GRENADES

A-26. Smoke grenades are used in urban combat to cover movement, deceive enemy forces as to the exact location of the attacking force, and for signaling. Smoke grenades should not be used in buildings and subterranean passages as they displace oxygen and protective masks do not filter the smoke or provide oxygen. Buildings in urban areas channel air currents, which influence emplacement of smoke grenades. One grenade should be used to test the airflow before employing additional grenades.

> **WARNING**
>
> **Avoid using smoke grenades in buildings and in subterranean passages.**

OPPOSING FORCES

A-27. OPFOR should be used as often as possible to enhance the realism of training, especially for FTXs, CFXs, and STXs. Thebe realistic, the OPFOR must have defined countertasks. These countertasks must support "Blue Force" training, and should prevent accidental contact between the Blue Force and the OPFOR. Leaders should develop a thorough control plan for OPFOR scenarios. This plan should describe uniform and vehicle markings, ammunition, safety instructions, controller guidance, and guidelines for handling enemy prisoners of war (EPW) and noncombatants. The ROE must clearly specify the limits of close combat between Soldiers.

- OPFOR heighten Soldiers' interest by giving them a real opponent. This fosters unit competition and realism, especially when the OPFOR is used in addition to TES and SESAM.
- OPFOR challenges leaders' ability to improvise. However, this free-play scenario should be monitored to ensure the unit meets the training objectives.
- OPFOR can use threat urban operations and doctrine to determine their course of action. If available, OPFOR uniforms, mock documents, overlays, and equipment add realism and intelligence to the scenario. Leaders conduct preliminary training for the OPFOR, which helps the OPFOR recreate the threat's urban tactics.
- Noncombatants should be integrated into training scenarios.

PRECISION MARKSMANSHIP TRAINING

A-28. Precision marksmanship training should be conducted before training at the UAC, CACTF, and the Shoot House, which accommodate training in target discrimination (FM 3-22.9). This is the best use of the unit's time and resources.

PHYSICAL READINESS TRAINING

A-29. Combat in urban areas places extraordinary physical demands on Soldiers. Therefore, physical readiness training, with emphasis on upper body development, is paramount to any urban operations training program.

VARIED SCENARIOS

A-30. Leaders must vary training scenarios to prevent redundant and predictable tactical behavior. Using predictable avenues of approach into the facility produces stereotypical tactical solutions.

BUILDING FACADE WALLS, WINDOWS, AND DOORS

A-31. The use of façade walls, windows and doors, in conjunction with SESAMS in the interior of some stations at the UAC, provides for more realistic training. This method of training reinforces the fact that walls, windows, and doors in most cases *do not* stop the penetration of service ammunition. Plastic wrap sheeting, tar paper, or some other type of material can be used to allow SESAMS penetration (Figure A-1).

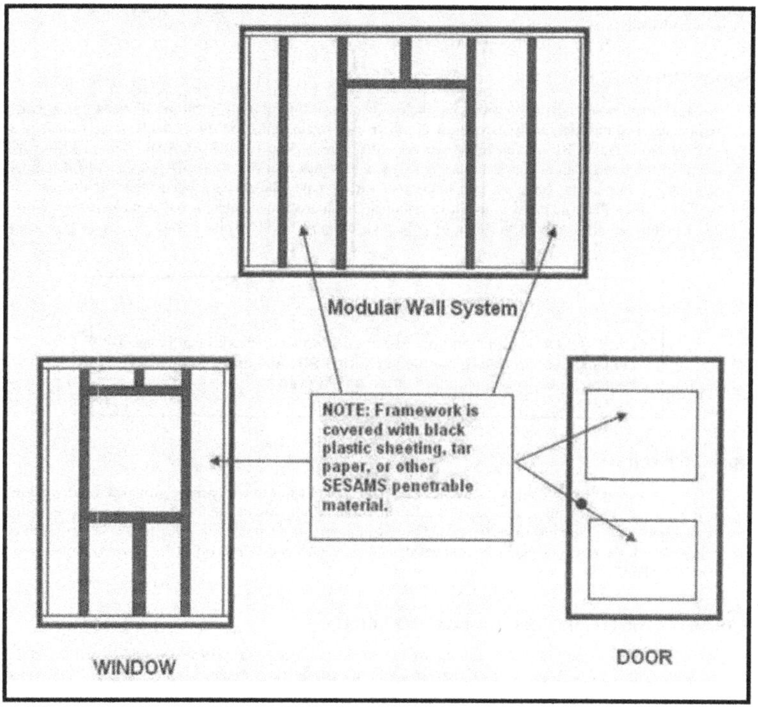

Figure A-1. Shoot-through walls and doors.

USE OF OLDER OR NONSTANDARD FACILITIES

A-32. Until installations receive funding for the UAC, SH, BF, and CACTF, they can use the older MOUT Assault Course (MAC) and Collective Training Facility (CTF) as described in the 2002 edition of TC 90-1, which this edition supersedes. Installations may also use nonstandard shoot houses and other nonstandard urban facilities. These facilities may be adapted as follows:

URBAN ASSAULT COURSE

A-33. Generally, the MAC can be used to train most of the UAC tasks. Of course, the MAC lacks targetry and instrumentation. Instead, units can record training with portable video cameras. During the AARs, they can play back the videos on VCRs hooked to TVs or on TV/VCR combos. OPFOR combatants will have SESAMS or MILES. People in civilian clothes can play the parts of the noncombatants.

SHOOT HOUSE

A-34. Many nonstandard shoot houses, such as tire houses, exist in different installations. Local range regulations and unit risk analyses will govern their use. Again, units can use portable video cameras for video capture and VCRs and television sets for video play back to facilitate an AAR. Both precision and nonprecision targets can be fabricated using E-type silhouettes on a stake, old BDUs/civilian clothes filled with straw, and so on. Simulate noncombatants with E-type silhouettes painted white or dummies in civilian clothes. Place balloons in the head and upper torso areas to simulate lethal shot placement. Where local policy permits, convert portions of collective training facilities into live-fire shoot houses using SRTA.

> **CAUTION**
>
> Units should ensure that they are in compliance with all local policies affecting live fire before attempting to use SRTA in an older collective training facility or nonstandard urban training site.

BREACH FACILITY

A-35. Existing demolition ranges may be used to conduct explosive breaching provided door, window, and wall units, or facsimiles, are available. Mechanical breaching can be trained in cantonment areas on the installation where building are designated for demolition, if such facilities are available. Regardless, whenever mechanical breacking is conducted, unit personnel must ensure that local safety policies are available, briefed, and complied with.

COMBINED ARMS COLLECTIVE TRAINING FACILITY

A-36. Generally the older CTF can be used to train the majority of tasks that would be trained in the CACTF. The CTF will not have any targetry and instrumentation. Again, units can use portable video cameras for video capture and VCRs and television sets for video play back to facilitate an AAR. Targets can be OPFOR with MILES. Role players in civilian clothes can be introduced to play the part of noncombatants. If armored vehicles are used, units should ensure that local range policies allow for their use. Older CTFs are normally either a 16 or 32 building design. Appropriate unit-level training, platoon, company, or battalion, should be planned based on the size of the facility.

Appendix B

Unit Training Matrix

Unit training plans are prepared IAW FM 7-1, which states that "Planning links the unit METL and the execution of battle-focused training.". The commander begins his planning with guidance from a higher echelon and an in-depth assessment of METL tasks (in this case, mission-essential urban tasks). Based on his own observations, feedback from his subordinates, and other sources the commander determines which tasks should be trained, which should be trained first, and how much time should be allotted for each. Tables B-1 through B-5 list training tasks by the WFF to be trained at each facility.

Table B-1. Army unit training matrix, brigade tasks.

Brigade Tasks	*SHOOT HOUSE* *Semiannual*	*BREACH HOUSE* *Semiannual*	*UAC* *Quarterly*	*CA/CTF* *Semiannual*	*CTC* **As Scheduled*	*SIMULATION* *Annual*
Develop Intelligence						
Perform Intelligence Operations					X	X
Perform S-2 Operations					X	X
Deploy/Conduct Maneuver						
Attack					X	X
Perform Air Assault					X	X
Defend					X	X
Lodgment					X	X
Exercise Command and Control						
Command and Control the Brigade					X	X
Maintain Communications					X	X
Coordinate Air Defense Operation					X	X
Conduct Noncombatant Evacuation					X	X
Develop a Media Plan					X	
Conduct Mediation and Negotiation					X	
Employ Firepower						
Execute Fires					X	X
Employ Air Support					X	X
Employ Attack Helicopters					X	X
Synchronize Fire Support					X	X
Synchronize CAS					X	X
Conduct Targeting Process					X	X
Protect the Force						
Perform Mobility/Survivability Operations					X	X
Provide Engineer Support					X	X
Coordinate CBRN Operations					X	X
Conduct Operations Security					X	X
Perform Sustainment						
Evacuate Casualties					X	X
Treat Casualties					X	X
Sustain the Force					X	X
Conduct a Civil-Military Operation					X	X
Provide Humanitarian Support					X	X

Table B-2. Army unit training matrix, battalion tasks.

Battalion Tasks	SHOOT HOUSE Semiannual	BREACH HOUSE Semiannual	UAC Quarterly	CA/CTF Semiannual	CTC As Scheduled	SIMULATION Annual
Develop Intelligence						
Conduct S-2 Operations				X	X	X
Deploy/Conduct Maneuver						
Attack a Built-Up Area				X	X	X
Plan for Urban Operation				X	X	X
Defend an Urban Area				X	X	X
Conduct an Infiltration/Exfiltration				X	X	X
Perform Cordon and Search Operations in an Urban Area				X	X	X
Plan for Urban Operations				X	X	X
Conduct Tactical Movement				X	X	X
Conduct Presence Operation in a Stability Environment				X	X	X
Defend an Urban Area				X	X	X
Traverse a Built-up Area				X	X	X
Conduct a Mobile Defense				X	X	X
Employ Firepower						
Integrate Fires and Effects Support				X	X	X
Establish the Fire Support Element				X	X	X
Protect the Force						
Conduct Mobility or Survivability Operations				X	X	X
Secure Civilians During Operation				X	X	X
Conduct Passive Air Defense Measures				X	X	X
Conduct Active Air Defense Measures				X	X	X
Employ Operations Security Measures				X	X	X
Integrate Air Defense Support				X	X	X
Perform Sustainment						
Conduct S-1 Operations				X	X	X
Conduct S-4 Operations				X	X	X
Conduct a Civil-Military Operation				X	X	X
Exercise Command and Control						
Conduct S-6 Operations				X	X	X
Conduct Noncombatant Evacuation				X	X	X
Conduct S-3 Operations				X	X	X
Conduct Mediation and Negotiations				X	X	X
Plan Roadblocks or Checkpoints				X	X	X
Establish Liaison				X	X	X

Table B-3. Army unit training matrix, company tasks.

Company Tasks	*SHOOT HOUSE Semiannual*	*BREACH HOUSE Semiannual*	*UAC Quarterly*	*CA/CTF Semiannual*
Develop Intelligence				
Conduct an Area or Zone Reconnaissance				X
Reconnoiter a Built-Up Area				X
Deploy/Conduct Maneuver				
Conduct Cordon and Search in a Built-Up Area				X
Attack a Built-Up Area				X
Conduct Presence Operations				X
Conduct Tactical Movement in a Built-Up Area				X
Defend in Urban Environment				X
Conduct an Infiltration (Co/Plt)				X
Conduct Operations with Armored or Mechanized Infantry Vehicles in an Urban Environment				X
Conduct Civil Disturbance Operations				X
Establish Checkpoints				X
Employ Firepower				
Integrate Direct Fires				X
Integrate Indirect Fires				X
Protect the Force				
Maintain Operations Security				X
Conduct Passive Air Defense Measures				X
Conduct Passive Air Defense Measures				X
Perform Sustainment				
Treat Casualties				X
Evacuate Casualties				X
Secure Civilians during Operations				X
Conduct Resupply Operations				
Process Captured Documents and				X
Equipment				X
Handle Enemy Prisoners of War				
Exercise Command and Control				
Prepare for Combat				X
Establish Communications				X
Conduct Consolidation and Reorganization				X
Conduct Negotiation				X
Conduct Troop-Leading Procedures				X

Table B-4. Army unit training matrix, platoon/squad tasks.

Platoon/Squad Tasks	*SHOOT HOUSE Semiannual*	*BREACH HOUSE Semiannual*	*UAC Quarterly*	*CA/CTF Semiannual*
Develop Intelligence				
Report Tactical Information	X		X	X
Conduct an Area or Zone Reconnaissance	X		X	X
Reconnoiter a Built-Up Area	X		X	X
Deploy/Conduct Maneuver				
Conduct Tactical Movement in a Built-Up Area	X		X	X
Assault a Building	X		X	X
Conduct a Strongpoint Defense of a Building				X
Conduct an Infiltration or Exfiltration				X
Conduct Operations with Armored or Mechanized Vehicles in an Urban Environment	X			X
Conduct a Presence Patrol	X			X
React to a Civil Disturbance				X
Establish a Checkpoint				X
Search a Building	X		X	X
Employ Firepower				
Employ Fire Support				X
Protect the Force				
Maintain Operations Security				X
Conduct Passive Air Defense Measures				X
Conduct Active Air Defense Measures				X
Perform Sustainment				
Secure Civilians During Operations	X		X	X
Perform Resupply Operations				X
Treat Casualties	X		X	X
Evacuate Casualties	X		X	X
Handle Enemy Prisoners of War				
Exercise Command and Control				
Conduct Troop-Leading Procedures	X		X	X
Establish Radio Communication	X		X	X
Conduct Consolidation and Reorganization	X		X	X
Prepare for Combat	X		X	X

Table B-5. Army unit training matrix, individual tasks/techniques.

Individual Tasks/Techniques	*SHOOT HOUSE* *Semiannual*	*BREACH HOUSE* *Semiannual*	*UAC* *Quarterly*	*CA/CTF* *Semiannual*
Skill Level 1				
Perform Movement Techniques in an Urban Environment, 071-326-0541	X	X	X	X
Prepare Positions for Individual and Crew-Served Weapons in an Urban Environment, 071-326-0550	X	X	X	X
Select Hasty Fighting Positions in an Urban Environment, 071-326-0557	X	X	X	X
Employ Hand Grenades, 071-325-4407	X	X	X	X
Engage Targets with an M16-Series or M16A2 Rifle, 071-311-2007	X	X	X	X
Execute a Mechanical Breach, 071-440-0027			X	X
Skill Level 2				
Conduct an Explosive Breach	X	X	X	X
Conduct a Breach, 071-440-0029	X	X	X	X
Prime Explosive Nonelectrically, 051-193-1003	X	X	X	X
Construct a Nonelectric Initiating Assembly with Modernized Demolition	X	X	X	X

Appendix C

Stability and Support Operations Unit Training Matrix

When assigned a stability or support mission, a well-trained unit must be able to rapidly shift its focus from war fighting to stability and support and also from stability and support to war fighting. During a stability or support operation, the unit performs numerous activities. Essentially, the unit accomplishes these activities through execution of tactical missions and tasks. While stability and support operations can occur anywhere, it is most probable that they will occur in an urban environment. Tables C-1 through C-4 list training tasks by the WFF to be trained at each facility.

Table C-1. Stability and support operations unit training matrix, brigade tasks.

Brigade Tasks	SHOOT HOUSE	BREACH HOUSE	UAC	CA/CTF	CTC	SIMULATION
Develop Intelligence						
Collect and Disseminate Information					X	X
Deploy/Conduct Maneuver						
Demonstrate a Show of Force					X	
Employ Psychological Operation					X	X
Employ Firepower						
Provide Fire Support					X	X
Protect the Force						
Protect the Force					X	
Perform Sustainment						
Restore Law and Order					X	
Exercise Command and Control						
Establish an Evacuation Control Center/Conduct Evacuation Control Center Operations					X	
Negotiate					X	
Perform Noncombatant Evacuation Operations					X	
Plan a Media Visit					X	
Develop Intelligence						
Collect and Disseminate Information				X	X	X

Table C-2. Stability and support operations unit training matrix, battalion tasks.

Battalion Tasks	*SHOOT HOUSE*	*BREACH HOUSE*	*UAC*	*CA/CTF*	*CTC*	*SIMULATION*
Deploy/Conduct Maneuver						
Interdict Smuggling Operations				X	X	
Disarm Belligerents				X	X	
Handle Captured Insurgents or Belligerents				X	X	
Employ Psychological Operation				X	X	
React to Civil Disturbance				X	X	
Employ Firepower						
Provide Fire Support				X	X	X
Protect the Force						
Protect the Force				X	X	
Restore Law and Order						
Identify and Process Detainees				X	X	
Deliver Supplies or Humanitarian Aid				X	X	
Control Civilian Movement				X	X	
Prepare Traffic Control Plan				X	X	
Exercise Command and Control						
Negotiate				X	X	
Perform Noncombatant Evacuation Operations				X	X	
Conduct Marshalling Force Operations				X	X	
Negotiate a Belligerent Force Checkpoint				X	X	
Plan a Media Visit				X	X	
Perform a Cordon and Search				X	X	
Coordinate and Monitor Convoy Security				X	X	

Table C-3. Stability and support operations unit training matrix, company tasks.

Company Tasks	*SHOOT HOUSE*	*BREACH HOUSE*	*UAC*	*CA/CTF*	*CTC*	*SIMULATION*
Develop Intelligence						
Collect and Disseminate Information				X	X	
Deploy/Conduct Maneuver						
Interdict Smuggling Operations				X	X	
Disarm Belligerents				X	X	
Handle Captured Insurgents or Belligerents				X	X	
Defend Convoy				X	X	
Apprehend/Detain Noncombatants				X	X	
Perform Cordon and Search				X	X	
Employ Firepower						
Provide Fire Support				X	X	
Protect the Force						
Establish a Checkpoint				X	X	
Operate a Checkpoint				X	X	
Protect the Force				X	X	
Perform Sustainment						
Process Captured Documents and Equipment				X	X	
Exercise Command and Control						
Negotiate a Belligerent Force Checkpoint				X	X	
Conduct Marshaling Force Operations				X	X	
Negotiate a Belligerent Force Checkpoint				X	X	
Plan a Media Visit				X	X	
Coordinate and Monitor Convoy Security				X	X	
Link Up with a Convoy				X	X	

Table C-4. Stability and support operations unit training matrix, platoon/squad tasks.

Platoon/Squad Tasks	*SHOOT HOUSE*	*BREACH HOUSE*	*UAC*	*CA/CTF*	*CTC*	*SIMULATION*
Deploy/Conduct Maneuver						
Disarm Belligerents				X	X	
Handle Captured Insurgents or Belligerents				X	X	
Apprehend/Detain Noncombatants				X	X	
Search a Building	X			X	X	
React to a Sniper			X	X	X	
Conduct Platoon Riot Control Formation				X	X	
Protect the Force						
Establish a Checkpoint				X	X	
Operate a Checkpoint				X	X	
Protect the Force				X	X	
Exercise Command and Control						
Negotiate a Belligerent Force Checkpoint				X	X	
Exercise Command and Control						
React to a Media Interview				X	X	

Appendix D

Leader Development Training

To be successful in combat, the Army must continually train to develop and maintain combat-ready Soldiers, leaders, and units that can perform assigned tasks to specific standards.

SIMULATIONS TRAINING

D-1. Commanders train their staffs to plan, coordinate, and synchronize their units' operations from initial mobilization and deployment. These operations include the full spectrum of Army operations; offense, defense, stability, and support. Current simulations lack adequate replication of urban scenarios, terrain, and weapons effects. The introduction of WARSIM 2000 and OneSAF should help to solve these simulations modeling deficiencies. For lower echelon units, battalion and below, constructive simulations complement collective battle staff training that occurs during virtual simulation similar to the close combat tactical trainer and live events. For echelons above battalion, it becomes difficult to use simulators or live maneuver to train. Studies to date indicate the crossover point for simulation versus simulator is at the battalion level. Some training events from battalion and below can be trained effectively using various types of simulators. Above battalion level, effective training almost has to be conducted with a simulation (Figure D-1).

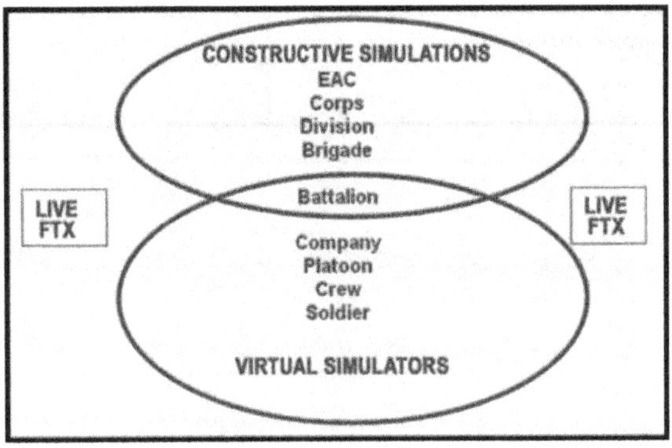

Figure D-1. Echelons and uses.

D-2. Simulations that support leader and battle staff training focus on collective tasks rather than individual leader tasks. Simulations, used in this context, rely on the human dynamics in the command setting and use real-time actions in a tactical environment with a unit's go-to-war systems. This process forces commander and staff interaction that provides feedback and helps refine the integration process. Battle staff training encourages team building, task coordination, and standardization of staff actions.

- Constructive simulations can provide commanders large areas for conducting training, access to sensitive areas through computer and mapping technology, and futuristically provide a much better level of fidelity in terms of urban operations terrain and scenario replication.
- Simulations can portray large, capable, and doctrinally correct OPFORs. All of the Soldiers portrayed in the units in a simulation do not have to be present in order for the training to be realistic. An attractive feature of simulations has always been the reduced staffing necessary to conduct this type of training exercise.
- Simulations can stress CPs and provide the realistic conditions under which commanders must make decisions. Brigade and below simulation (BBS) and corp battle simulation (CBS) are well known for the stressful environments created during an exercise. The stress is realistic and can assist in conditioning the training audience before CTC rotations.
- Simulations allow different units to train under the same conditions and to the same standards. This technique is a unique feature of simulations as the exact weather, opposing force organization, location of units, and equipment can be used to train many different units on the same terrain with the same conditions.
- The size and number of personnel in a battle staff vary depending on the echelon and other demands unique to that commander. However, the WFF are represented by the following staff officers: intelligence, operations, fire support, communications, logistics, air defense, and engineer.
- Characteristics and benefits of simulations used to support battle staff training follow:
  - -- Build and sustain staff teamwork.
  - -- Build and sustain effective and efficient organizations.
  - -- Provide training and validation of command and staff procedures (SOPs, TACSOPs, FSOPs).
  - -- Provide proficiency to the individual staff officer, while building proficiency in a fully integrated staff.
  - -- Allow practice of battle drills and tactical missions (movement to contact, hasty defense).
  - -- Allow commanders to determine if their staff and subordinate commanders understand the commander's intent and concept of operations.

D-3. The major development efforts for FAMSIM is WARSIM 2000. WARSIM 2000, in conjunction with OneSAF, exploits new technologies to enable commanders and staffs at battalion through EAC to train in a realistic simulation environment (Table D-1). The increased realism of WARSIM 2000 over existing models allows units to use Army Battle Command Systems (ABCS) to synchronize across the complete range of operations. WARSIM 2000's design allows warfighting CPs to use organic ABCS equipment to interact with the simulation from training sites in the field, as well as from simulation centers. WARSIM 2000 will also be capable of showing a joint and combined environment across the operational continuum.

Table D-1. Urban operations simulations crosswalk.

CAMTF Training Tasks	*WARSIM*	*OneSAF*
Corps	X	
Division	X	
Brigade		X
Battalion		X
Company		X
Platoon		X

D-4. Future simulation capabilities simultaneously integrate live, virtual, and constructive environments. Constructive simulation is a viable solution for battalion and above training and the federation of WARSIM and OneSAF models should alleviate the following training simulation problems:

- Eliminate the need for swivel chair interface.
- Allows focus on subordinate units fighting the fight.
- Digital and analog feeds to C4I systems including FBCB2.
- May provide for distance learning opportunities.
- Facilitates simultaneous multi-echelon urban operations training

D-5. At company and below, the best training solution is live and virtual. This is accomplished by linking the Close Combat Tactical Trainer (CCTT) to OneSAF and VLET to Engagement Skills Trainer (EST) technology. Entity-based models like OneSAF can allow the leader to fight from a TOC or vehicle, if linked to virtual or live environments. It also has the potential for distance learning.

PHASES OF TRAINING

D-6. The Army Urban Operations Training Strategy consists of initial, intermediate, and culminating stages. These phases of training, shown in Figure D-2, describe the sequence in which urban operations training tasks should be conducted.

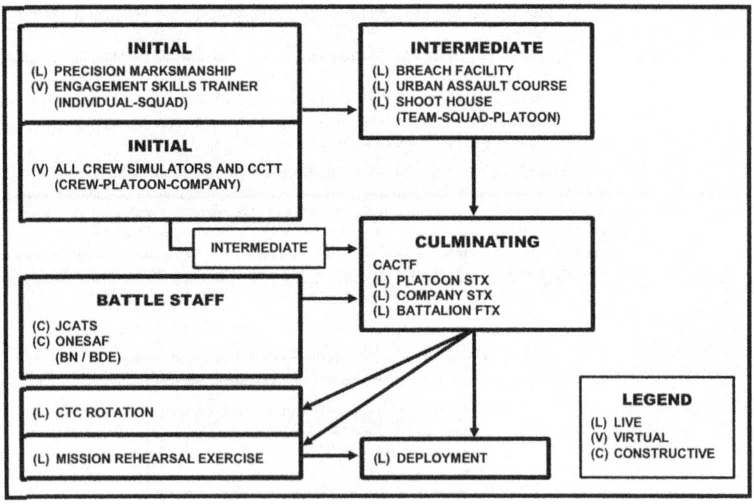

Figure D-2. Phases in the Live, Virtual, and Constructive Training Strategy.

INITIAL PHASE

D-7. The initial phase of training covers those individual and collective tasks that Soldiers and leaders should be proficient at before arriving at any of the urban operations training sites. They consist primarily of basic marksmanship skills and Soldier tasks needed to perform any tactical operation.

INTERMEDIATE PHASE

D-8. The intermediate phase consists of training conducted at the breach facility, shoot house, and the urban assault course. Urban operations training tasks should be conducted here before advancing to the Combined Arms Collective Training Facility with the exception of Aviation, Armor, and other branches that have no practical use for the Shoot House, Breach Facility, or Urban Assault Course.

CULMINATING PHASE

D-9. The combined arms collective training facility is the culminating phase of urban operations training and focuses on multiechelon collective skills for all branches of the combined arms team. Urban operations leader training currently exists in our leader professional development courses. Units have developed internal urban warfare training programs and TRADOC schools have assembled technical mobile training teams (MTTs) for marksmanship, combatives, and breaching techniques.

EDUCATION PHASE

- Urban warfare reading/video list.
- Urban operations equipment.
- Review of doctrinal publications (FM 3-21.10, FM 3-06.11, for Infantry) and corresponding publications for other branches.
- Small-unit urban movement techniques and considerations.

PLANNING AND EXECUTION PHASE

- Prepare company OPORD or WFF-related products.
- Brief platoon/company OPORD/WFF-related products to commander or staff representative.
- Participate as briefer in urban environment TEWT.

SUSTAINMENT PHASE

- Participate in company/platoon TEWT or other urban operations event.
- Execute/facilitate a company or platoon TEWT.

EXAMPLE LEADER'S COURSE POI

D-10. An example urban leaders' course POI, which is intended to provide unit trainers with a tailorable urban operations training program, is shown in Table D-2. It is a *train-the-trainer* course for team leaders through platoon leaders to train tactics, techniques, and procedures of advanced urban operations skills. The length and focus of the POI can be tailored to support the commander's training objectives and unit METL. The target audience is newly arrived, or soon to be, small unit leaders. The recommended course length is about two weeks or 120 to 140 hours of instruction.

REALISTIC TRAINING OFF FEDERAL FACILITIES

D-11. Realistic training for urban operations is critical to force readiness. Training opportunities that are both realistic and challenging may be unavailable at federal installations, or at facilities specifically designed for urban operations training. Training in off-post civilian urban settings may become necessary in order to support readiness and training requirements.

D-12. The Department of Defense has developed the following SOP for planning and approving such events. The SOP assists in balancing the needs of the military training and of the civilian community/ies to be used for realistic urban training events.

Table D-2. POI for urban leader's train-the-trainer course.

Course Overview	• Introduction	
Precision Marksmanship	• Firing techniques. • Target discrimination.	
Breaching Techniques	• Explosive. • Ballistic. • Mechanical.	

Table D-2. POI for urban leader's train-the-trainer course (continued).

Offensive Operations	• Planning considerations	-- METT-TC. -- Snipers. -- CASEVAC. -- Communications. -- ROE.
	• Platoon attack a built-up area (task organization)	-- Assault (isolate, enter, clear). -- Rooms, hallway, stairs, superstructure, subsurface. -- Markings. -- Movement techniques. -- Streets. -- Alley ways. -- Obstacles. -- Other danger areas. -- Open areas. -- Roof tops.
	• Combined arms	-- Armor/mechanized platforms. -- Aviation assets. -- Artillery. -- Engineer.
	• Limited visibility	
Defensive Operations	• Planning considerations	-- METT-TC. -- Snipers. -- CASEVAC. -- Communications. -- ROE.
	• Characteristics of urban areas	
	• Hasty defense	-- Occupation and preparation of positions. -- Hasty firing positions. -- Improving the defense.
	• Deliberate defense, including priorities of work and other considerations	
	• Defensive plan at platoon level.	-- Defense of a strongpoint. -- Defense against armor.
	• Combined arms	-- Armor/mechanized. -- Aviation assets. -- Artillery. -- Engineer.
	• Limited visibility • Fundamental combat skills	

Table D-2. POI for urban leader's train-the-trainer course (continued).

Stability and Support Operations	• Principles of stability and support.	
	• Stability and support activities	-- NEO. -- Humanitarian assistance. -- Disaster relief. -- Show of force. -- Attack and raids.
	• Rules of engagement application • Convoy operations • Checkpoints • Fire support • Sniper employment/counter sniper measures • Civil disturbance	
Urban Training Facilities	• Breach facility • Urban assault course • Shoot house • CACTF	
New Technologies	• Rifle launch entry munition	
	• Unmanned vehicles	-- UAV. -- UGV. -- Ladders. -- Night vision devices.

STANDING OPERATING PROCEDURES FOR PLANNING AND APPROVING REALISTIC URBAN TRAINING

1. **REFERENCES**.

   10 USC § 167, *Unified Combatant Command for Special Operations Forces*.
   10 USC §§ 3013, 5013, 8013.
   Department of Defense Directive 5111.10.
   Unified Command Plan 99.

2. **PURPOSE**. This SOP establishes uniform planning and approval procedures for realistic urban training (RUT) events.

3. **APPLICABILITY**. This SOP applies to all RUT events in civilian urban settings within the United States, its territories, and its possessions; and to Active or Reserve Forces, including National Guard Forces in Federal service. This policy does not apply to National Guard Forces under in-state status (Title 32), or to aviation operations that do not require an exemption from standard Federal Aviation Regulations. The training event should not be located on federal state, local, or private installations or facilities.

4. **DEFINITIONS**. For the purposes of this SOP, the following definitions apply:

a. **Urban Operations. This term** includes (see FM 3-06) all military actions planned and conducted on a topographical complex, and on adjacent natural terrain, where man-made construction is the dominant feature. It includes combat in towns and cities, including house-to-house and street-by-street fighting.

b. **Realistic Urban Training.** RUT is high-intensity, close-quarter battle training using live or nonlethal fires; demolitions or explosives such as breaching charges; and air and naval supporting platforms at the objective, in ***civilian urban settings***.

c. **Civilian Urban Setting.** A civilian urban setting is any area(s) located in or near a civilian community. Civilian urban settings will not be developed or established for training events, especially where training activities would affect civilian residences or commercial areas due to noise, traffic, and so on.

5. **PROCEDURES**. The following procedures apply to the planning and approval of RUT in civilian urban settings:

a. The combatant commander or the service ensures compliance with this SOP. Approval authority for RUT events may be delegated to a general or flag officer within the chain of command, or to the combatant common service conducting the training.

b. The approval authority reviews proposed training events to determine whether they are necessary and appropriate for unit readiness. Training must support unit METs and JMETs.

c. On approval of the proposed training, and at least 14 days before training starts, the combatant common service notifies the **SECR/PF** through the CJCS (by way of message) of the approved training. At a minimum, this notice includes the following:

- Name and location of the unit conducting training.
- Scope of training and activities planned such as live fire, breaches, or use of aircraft.
- Start and end dates.
- Training location.
- A statement that all necessary coordination has been accomplished IAW this SOP. The statement should include a list of the contacted official's names and their positions.

d. **Coordination with Civilian Officials**.

- While planning the training event, the commander coordinates with and obtains written approval to conduct the event from the appropriate federal, state, and local civilian authorities. Commanders will consult their legal and public affairs officers to identify the civilian officials, for example the mayor, borough chief, county commissioners or supervisors, with whom they must coordinate the training event. To do this, they consult local and Federal law enforcement agencies. At a minimum, they must consult a senior official with responsibility for each affected civilian urban setting.
- During coordination, the commander covers the details of the proposed training activities, such as whether the event involves live fire, rotary- or fixed-wing aircraft, or close quarters battle training; and whether the exercises will be conducted during the day, night, or both: and so on. It will also list actions to ensure public and participant safety, proposed public or resident notification and outreach procedures, and any other issues that might cause public interest or concern.
- As much as possible, the commander briefs civilian officials in person on the exercise concept.

- The US congressional representative or staff for the training district, and the US senators for the state, or their staffs, are advised of the proposed training event and given whatever information they request. Neither written approval nor concurrence from the representative or senators is required. However, if these officials raise objections or issues, and if those issues cannot be resolved, they are immediately elevated to the Joint Staff, J3, and the service or combatant commander point of contact.
- For each training site, license agreements with civilian or government property owners or managers are executed. Agreements address liability and conditions of acceptable usage of property. Property owners are informed of the nature of the training including written notification as to whether the activities will include use of breaching charges and, if so, how much; or live ammunition and, if so, which weapons and what ammunition. This notice will also state *if* rotary or fixed-wing aircraft will be used *and* whether the exercises will be conducted during the day, night, or both. Before the exercise can begin, all license agreements must be reviewed for legality. Copies of the license agreements will be provided to the senior civilian officials responsible for training area.
- A record of coordination activities is maintained for each training event. The record includes the names and positions of the officials *who* coordinate and approve the event. It also provides officials with a summary of the events such as date, time, and location of the training; types of activities; and planned public notification and outreach. A follow-on letter is distributed to the senior civilian officials responsible for the affected civilian urban settings. The letter confirms the training event *and* lists the names of the civilian officials who plan the training event.
- The unit may schedule recurring training events at a particular area or location by submitting a single MOA to the authorizing civilian officials. The terms of the MOA must comply with the requirements of this SOP. The service or the combatant commander reviews the MOA, which authorizes the recurring training, to ensure it complies with the procedures. Once the commander approves the MOA, the unit may conduct individual training events under it without further Joint Staff or OSD review. However, services and combatant commanders notify SO/LIC *and* the Joint Staff 14 days *before* the training in this MOA begins. Thereafter, services and combatant commanders must review these MOAs periodically to ensure *that* they continue to fulfil the intent.
- *If* nondisclosure agreements are required, command legal representatives must review them. Nondisclosure statements should clearly explain *that* the agreement does not preclude persons from informing others of general, nonsensitive information about the training exercise.
- Depending on OPSEC and safety considerations, commanders must adhere to the recommendations of civilian officials regarding community notification, outreach, and press procedures. Community notification and outreach procedures may include door-to-door notification by the local law enforcement agency. Commanders will advise civilian officials when they cannot follow their recommendations.

e. Other instructions include--

- Legal and PA representatives attend final planning meetings, and they remain on-site throughout all training activities.
- The on-site PAO handles all press queries after coordinating with the on-scene commander.
- All requested training must follow applicable statutes, law, DOD directives, other guidance, and implementing service directives.

Glossary

Section I. ACRONYMS AND ABBREVIATIONS

1SG first sergeant

A

AA assembly area
AAR after-action review/report
ABCS Army Battle Command System
ACU Army combat uniform
ADA air defense artillery
ANCOC advanced noncommissioned officer course
AO area of operation
AR Army Reserve
AT4 a shoulder-fired munition, formerly called a "light antiarmor weapon"
atk attack
ATWESS antitank weapon effect signature simulator
AVCATT armored vehicle combined arms tactical trainer

B

BBS brigade and below simulation
BCT brigade combat team
bde brigade
BDU (obsolete; *see* ACU)
BF Breach Facility
bn battalion
BNCOC basic noncommissioned officer course
BOLC Basic Officer Leadership Course
BOS (obsolete; *see* WFF)
BP battle position
BSA brigade support area
BSX battle simulation exercise

C

C2 command and control
CACTF Combined Arms Collective Training Facility
CASEVAC casualty evacuation
CBRN chemical, biological, radiological, nuclear
CBS corps battle simulation
CCMCK close combat mission capability kit
CCOC Captain's Career Officer Course
CCTT close combat tactical trainer
CCTV closed-circuit television
CFX command field exercise
CMTC Citizens Military Training Corps
co company
commo communications
CP command post
CPX command post exercise
CTC Combat Training Center
CTF collective training facility

D

DES demolition effects simulator
det cord detonation cord
div division
DLIC detachment left in contact
DS Digital Training Management System
DTMS direct support

E

EA engagement area
EAC echelons above corps
ENCATT engineer combined arms tactical trainer
engr engineer
EPW enemy prisoners of war
EST Engagement Skills Trainer

F

FAMSIM family of simulations
FO forward observer
FRAGO fragmentary order
FSCATT fire support combined arms tactical trainer
FSOP field standing operating procedures
FTX field training exercise

G

GREM	grenade rifle entry munition
GSR	ground surveillance radar

H

HC	herachloroethane zinc
HEAT	high-explosive antitank

I

IAW	in accordance with
Inf	Infantry

J

JANUS	Joint Army-Navy Uniform Simulation
JCATS	Joint Conflict and Tactical Simulation
JMET	joint mission-essential task
JRTC	Joint Readiness Training Center
JTF	joint task force
JTX	joint training exercise

L

LAW	light antiarmor weapon
LFX	live-fire exercise
LMG	light machine gun

M

MAC	MOUT assault course
MACOM	major command
MAPEX	map exercise
M/C/S	mobility/countermobility/ survivability
MET	mission-essential task
METL	mission-essential task list
METT-TC	mission, enemy, terrain, troops, time available, and civilians
MI	military intelligence
MILES	Multiple Integrated Laser-Engagement System
MOPP	mission-oriented protective posture
MPAT	multipurpose antitank (ammunition)
MPAT-OR	multipurpose antitank-obstacle reducing
MRE	meals ready to eat
MTT	mobile training team

N

NBC	(obsolete; *see* CBRN)
NEO	noncombatant evacuation operations
NLT	not later than
NTC	National Training Center

O

OneSAF	one semiautomated force (software used to create modeling and simulation programs with military applications)
op	operation
OPFOR	opposing force
OPORD	operation order
OPSEC	operations security

P

plt	platoon
POI	program of instruction
PPP	power projection platform
PSP	power support platform

R

ROE	rules of engagement
ROI	rules of interaction
RUT	realistic urban training

S

S-1	personnel staff officer (adjutant)
S-2	intelligence staff officer
S-3	operations staff officer
S-4	logistics staff officer
S-5	civil affairs officer
S-6	command, control, communications and computer operations (C4 operations) officer
SAAF	small-arms alignment fixture
SBCT	Stryker brigade combat team
SDZ	surface danger zone
SEP	system enhancement program: a version of the M1A2 machine gun with enhanced digital command and control and second-generation forward-looking infrared

SESAMS	Special Effects Small Arms Marking System
SH	Shoot House
SOI	signal operation instructions
SOP	standing operating procedure
sqd	squad
SRTA	short-range training ammunition
STAFFEX	staff exercise
STX	situational training exercise

T

T&EO	training and evaluation outline
TACSOP	tactical standing operating procedure
TES	Tactical Engagement System
TEWT	tactical exercise without troops
TF	task force
TOW	tube-launched, optically tracked, wire-guided (missile)
TTP	tactics, techniques, and procedures

U

UAC	urban assault course
UO	urban operations

V

VCR	videocassette recorder
VLET	virtual leader effects trainer

W

WARSIM	warfighting simulation
WFF	warfighting function
WFX	warfighting exercise

X

XO	executive officer

Section II. TERMS

bustle rack	storage rack mounted on turrets of some armored vehicles
entity-based model	a representation of an actual or conceptual system that includes mathematical, logistical, or computer simulation
warfighting functions	group of tasks and systems (people, organization, information, and processes) united by a common purpose that commanders use to accomplish missions and training objectives. The warfighting functions are intelligence, movement and maneuver, fire support, protection, sustainment, and command and control. These warfighting functions replace the battlefield operating systems (BOSs).

References

SOURCES USED/DOCUMENTS NEEDED

These are the sources quoted or paraphrased in this publication. All of them are needed by and must be available to the intended user.

DOD Directive

DOD Directive 2311.01E *DOD Law of War Program.* 9 May 2006.

Field Manuals

FM 3-06	*Urban Operations.* 26 October 2006.
FM 3-06.11	*Combined Arms Operations in Urban Terrain.* 28 February 2002.
FM 3-20.12	*Tank Gunnery (Abrams).* 15 August 2005.
FM 3-21.8	*The Infantry Platoon and Squad.* 28 March 2007.
FM 3-21.9	*SBCT Infantry Rifle Platoon and Squad.* 2 December 2002.
FM 3-21.10	*The Infantry Rifle Company.* 27 July 2006.
FM 3-21.11	*The SBCT Infantry Rifle Company.* 23 January 2003.
FM 3-21.20	*The Infantry Battalion.* 13 December 2006.
FM 3-21.21	*The Stryker Brigade Combat Team Infantry Battalion.* 8 April 2003.
FM 3-21.71	*Mechanized Infantry Platoon and Squad (Bradley).* 20 August 2002.
FM 3-22.9	*Rifle Marksmanship M16A1, M16A2/3, M16A4, and M4 Carbine.* 24 April 2003.
FM 7-0	*Training the Force.* 22 October 2002.
FM 7-1	*Battle Focused Training.* 15 September 2003.
FM 25-4	*How to Conduct Training Exercises.* 10 September 1984.

Soldier Training Publications

STP 21-1-SMCT	*Soldier's Manual of Common Tasks Skill Level 1.* 14 December 2007.
STP 21-24-SMCT	*Soldier's Manual of Common Tasks (SMCT) Warrior Leader Skill Levels 2-4.* 2 October 2006.

Training Circular

TC 7-98-1	*Stability and Support Operations Training Support Package.* 5 June 1997.

Index

Made in United States
Orlando, FL
04 November 2026

90218350R00154